BOARD REVIEW SERIES

Genetics

D0851675

Genetics

Ron W. Dudek PhD

Full Professor
East Carolina University
Brody School of Medicine
Department of Anatomy and Cell Biology
Greenville, NC 27858

Wolters Kluwer | Lippincott Williams & Wilkins
Health

Philadelphia · Baltimore · New York · London
Buenos Aires · Hong Kong · Sydney · Tokyo

Acquisitions Editor: Susan Rhyner
Managing Editors: Stacey Sebring and Jennifer Verbiar
Marketing Manager: Jennifer Kuklinski
Design Coordination: Holly Reid McLaughlin
Interior Designer: Karen Quigley
Cover Designer: Larry Didona
Compositor: Aptara

QU
18.2
D845g
2009
c.2

9 8 7 6 5 4 3 2 1

Library of Congress Cataloging-in-Publication Data

Dudek, Ronald W., 1950-
  Genetics / Ron W. Dudek. — 1st ed.
      p. ; cm. — (Board review series)
  Includes bibliographical references and index.
  ISBN 978-0-7817-9994-2 (alk. paper)
  1. Genetics—Examinations, questions, etc. I. Title. II. Series: Board
review series.
  [DNLM: 1. Genetic Phenomena—Examination Questions. 2. Genetic Diseases,
Inborn—Examination Questions. QU 18.2 D845g 2009]
  QH440.3.D83 2009
  576.5078—dc22

                                                    2009000283

DISCLAIMER

Care has been taken to confirm the accuracy of the information present and to describe generally accepted practices. However, the authors, editors, and publisher are not responsible for errors or omissions or for any consequences from application of the information in this book and make no warranty, expressed or implied, with respect to the currency, completeness, or accuracy of the contents of the publication. Application of this information in a particular situation remains the professional responsibility of the practitioner; the clinical treatments described and recommended may not be considered absolute and universal recommendations.

The authors, editors, and publisher have exerted every effort to ensure that drug selection and dosage set forth in this text are in accordance with the current recommendations and practice at the time of publication. However, in view of ongoing research, changes in government regulations, and the constant flow of information relating to drug therapy and drug reactions, the reader is urged to check the package insert for each drug for any change in indications and dosage and for added warnings and precautions. This is particularly important when the recommended agent is a new or infrequently employed drug.

Some drugs and medical devices presented in this publication have Food and Drug Administration (FDA) clearance for limited use in restricted research settings. It is the responsibility of the health care provider to ascertain the FDA status of each drug or device planned for use in their clinical practice.

To purchase additional copies of this book, call our customer service department at **(800) 638-3030** or fax orders to **(301) 223-2320**. International customers should call **(301) 223-2300**.

Visit Lippincott Williams & Wilkins on the Internet: http://www.lww.com. Lippincott Williams & Wilkins customer service representatives are available from 8:30 am to 6:00 pm, EST.

Acknowledgments

I would like to express my thanks to Betty Sun for having the confidence in me so that I could write *BRS Genetics*. My working relationship with Betty Sun has extended over many years and many book projects. It was my pleasure and privilege to work with Betty Sun, who always brought wise counsel, keen insight, and common sense to the table.

I would also like to thank all the LWW staff who played a role in *BRS Genetics*, including Kathleen Scogna, Stacey Sebring, Jen Clements, Jenn Verbiar, Jennifer Kuklinski, and Sally Glover.

Acknowledgments

I would like to express my thanks to Betty Sun for having the confidence in me so that I could write SAS Genetics. My working relationship with Betty Sun has extended over many years and many book projects. It was my pleasure and privilege to work with Betty Sun, who always brought wise counsel, keen insight, and common sense to the table.

I would also like to thank all the LWW staff who played a role in SAS Genetics, including Kathleen Scogna, Stacey Sebring, Jen Clements, Jennifer Verbiar, Jennifer Kuklinski, and Sally Glover.

Preface

Since many US medical schools are unable to find adequate time in the curriculum for an in-depth genetics course, medical students find themselves in a less than advantageous position when reviewing genetics for the USMLE Step 1. A brief visit to any medical bookstore will reveal that there are about six excellent genetics textbooks that cover basic genetics and modern molecular genetic advancements. However good these books are, they are not designed for a review process under the time constraints that medical students face when preparing for the USMLE Step 1. Consequently, I wrote BRS Genetics with the goal of placing the student in a strategic position to review genetics in a reasonable time period and most importantly to answer all the Genetics questions that would likely appear on the USMLE Step 1. *BRS Genetics* expands many of the topics that have been included in *High Yield Genetics* for the student that needs a little more background and wants a little more depth. In addition, *BRS Genetics* has test questions after each chapter and a comprehensive exam that should serve the student well not only for USMLE Step 1 but also in their coursework.

Discussions concerning the preparation for the USMLE Step 1 usually include mention of the "big three": pathology, pharmacology, and physiology. For many USMLE Step 1 clinical case style of questions, these three disciplines coordinate very nicely to present a clinical case and then ask a mechanistic question as to WHY something is observed or HOW a specific drug treatment works. The "big three" has become a perfect triad for USMLE Step 1 preparation.

However, in the future, I think that a "new big three" will develop: embryology, genetics, and molecular biology. With the completion of the Human Genome Project and now the advancement of genome mapping for every individual, the future of medicine will revolve around the elucidation of the genetics of birth defects and other human diseases spearheaded by molecular biology techniques. Exactly when the "new big three" will have significant representation on the USMLE Step 1 is impossible for me to predict. However, when this does occur, the BRS series will be strategically placed to serve its customers with three superb publications: *BRS Embryology, BRS Genetics, BRS Biochemistry and Molecular Biology* (by Swanson, Kim, and Gluckman), and High Yield Cell and Molecular Biology. These books are well integrated, have minimal overlap, and are updated with the latest information.

More than any other field of medicine in the future, genetics and molecular biology will be the engines that drive new breakthroughs and new information that are relevant to clinical practice. This will require that *BRS Embryology, BRS Genetics, BRS Biochemistry and Molecular Biology* (by Swanson, Kim, and Gluckman), and High Yield Cell and Molecular Biology be routinely updated with new clinically relevant information. For this, I rely on my readers to e-mail me suggestions, comments, and new ideas for future editions. Please e-mail me at dudekr@ecu.edu

Dr. Ron W. Dudek

Contents

5.   UNIPARENTAL DISOMY AND REPEAT MUTATIONS   45

6.   MITOCHONDRIAL INHERITANCE   53

7.   MULTIFACTORIAL INHERITED DISORDERS   63

16.   GENETICS OF CANCER   169

17.   GENETIC SCREENING   185

18.   CONSANGUINITY   197

chapter 1

The Human Nuclear Genome

I. GENERAL FEATURES (Figure 1-1)

A. The human genome refers to the haploid set of chromosomes (nuclear plus mitochondrial), which is divided into the very complex **nuclear genome** and the relatively simple **mitochondrial genome** (discussed in Chapter 6).

B. The human nuclear genome consists of 24 different chromosomes (22 autosomes; X and Y sex chromosomes). The human nuclear genome codes for ≈**30,000 genes** (precise number is uncertain) which make up ≈**2% of human nuclear genome.**

C. There are ≈**27,000 protein-coding genes** (i.e., they follow the central dogma of molecular biology: DNA transcribes RNA → mRNA translates protein).

D. There are ≈**3,000 RNA-coding genes** (i.e., they do not follow the central dogma of molecular biology: DNA transcribes RNA → RNA is *not* translated into protein).

E. The fact that the ≈30,000 genes make up only ≈2% of the human nuclear genome means that ≈**2% of the human nuclear genome consists of coding DNA** and ≈**98% of the human nuclear genome consists of noncoding DNA.**

F. When the **Human Genome Project** identified ≈30,000 genes, it was somewhat of a surprise to find such a low number especially when compared to the genome of the roundworm (*Caenorhabditis elegans*). The roundworm genome codes for ≈19,100 protein-coding genes and >1,000 RNA-coding genes. This means that there is no correspondence between biological complexity of a species and the number of protein-coding genes and RNA-coding genes **(i.e., biological complexity ≠ amount of coding DNA)**. However, there is correspondence between biological complexity of a species and the amount of noncoding DNA **(i.e., biological complexity = amount of noncoding DNA)**.

G. In order to fully understand how heritable traits (both normal and disease related) are passed down, it is important to understand three aspects of the human nuclear genome, which include the following:
  1. **Protein-coding genes.** For decades, protein-coding genes were enshrined as the sole repository of heritable traits. A mutation in a protein-coding gene caused the formation of an abnormal protein and hence an altered trait or disease. Today, we know that protein-coding genes are not the sole repository of heritable traits and that the situation is more complicated.
  2. **RNA-coding genes.** RNA-coding genes produce **active RNAs** that can profoundly alter normal gene expression and hence produce an altered trait or disease.

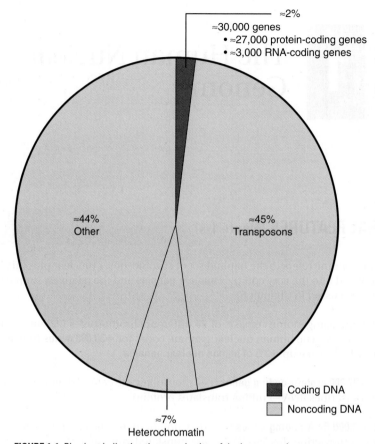

≈2%

≈30,000 genes
• ≈27,000 protein-coding genes
• ≈3,000 RNA-coding genes

≈44%
Other

≈45%
Transposons

Coding DNA

Noncoding DNA

≈7%
Heterochromatin

FIGURE 1-1. Pie chart indicating the organization of the human nuclear genome.

3. **Epigenetic control.** Epigenetic control involves **chemical modification of DNA** (e.g., methylation) and **chemical modification of histones** (e.g., acetylation, phosphorylation, addition of ubiquitin), both of which can profoundly alter normal gene expression and hence produce an altered trait or disease.

II. PROTEIN-CODING GENES

A. **Size.** The size of protein-coding genes varies considerably from the 1.7 kb insulin gene → 45 kb LDL receptor gene → 2,400 kb dystrophin gene.

B. **Exon-Intron Organization.** Exons (expression sequences) are coding regions of a gene with an average size of <200 bp. Introns (intervening sequences) are noncoding regions of a gene with a huge variation in size. A small number of human genes (generally small genes <10 kb) consists only of exons (i.e., no introns). However, most genes are composed of exons and introns. There is a direct correlation between gene size and intron size (i.e., large genes tend to have large introns).

C. **Repetitive DNA Sequences.** Repetitive DNA sequences may be found in both exons and introns.

D. **Classic Gene Family.** A classic gene family is a group of genes that exhibit a high degree of sequence homology over most of the gene length.

E. Gene Superfamily. A gene superfamily is a group of genes that exhibit a low degree of sequence homology over most of the gene length. However, there is relatedness in the protein function and structure. Examples of gene superfamilies include the immunoglobulin superfamily, globin superfamily, and the G-protein receptor superfamily.

F. Organization of Genes in Gene Families.
   1. **Single cluster.** Genes are organized as a **tandem repeated array; close clustering** (where the genes are controlled by a single expression control locus); and **compound clustering** (where related and unrelated genes are clustered) all on a single chromosome.
   2. **Dispersed.** Genes are organized in a dispersed fashion at two or more different chromosome locations all on a single chromosome.
   3. **Multiple clusters.** Genes are organized in multiple clusters at various chromosome locations and on different chromosomes.

G. Unprocessed Pseudogenes, Truncated Genes, Internal Gene Fragments.
   1. Gene families are typically characterized by the presence of unprocessed pseudogenes (i.e., defective copies of genes that are not transcribed into mRNA); truncated genes (i.e., portions of genes lacking 5′ or 3′ ends); or internal gene fragments (i.e., internal portions of genes), which are formed by **tandem gene duplication.**
   2. In humans, there is strong selection pressure to maintain the sequence of important genes. So, in order to propagate evolutionary changes, there is a need for gene duplication.
   3. The surplus duplicated genes can diverge rapidly, acquire mutations, and either degenerate into nonfunctional pseudogenes or mutate to produce a functional protein that is evolutionary advantageous.

H. Processed Pseudogenes. Processed pseudogenes are transcribed into mRNA, converted to cDNA by reverse transcriptase, and then the cDNA is integrated into a chromosome. A processed pseudogene is typically not expressed as protein because it lacks a promoter sequence.

I. Retrogene. A retrogene is a processed pseudogene where the cDNA integrates into a chromosome near a promoter sequence by chance. If this happens, then the processed pseudogene will express protein. If selection pressure ensures the continued expression of the processed pseudogene, then the processed pseudogene is considered a **retrogene.**

J. The Human Proteome. The Human Genome Project has allowed the construction of a number of databases based on the DNA sequences that are shared by multiple proteins and indicate common functions. These databases have been organized into various protein families, protein domains, molecular function, and biological process.
   1. **Protein families.** The largest protein family consists of **rhodopsin-like G protein coupled receptors.** The second largest protein family consists of **protein kinases.**
   2. **Protein domains.** The most abundant protein domain is a **zinc finger C2H2 type** domain.
   3. **Molecular functions.** The most common molecular function of a protein is **ligand binding.** The second most common molecular function of a protein is **enzymatic.**
   4. **Biological processes.** The most common biological process that proteins are involved in is **protein metabolism.** The second most common biological process that proteins are involved in is **DNA, RNA,** and other **metabolic processes.**

III. RNA-CODING GENES

A. 45S and 5S Ribosomal RNA (rRNA) Genes.
   1. The rRNA genes encode for **rRNAs** that are used in **protein synthesis.**
   2. The nucleolar organizing regions are the portions of the short arm of five pairs of chromosomes (i.e., 13, 14, 15, 21, and 22) that contain about **200 copies** of rRNA genes, which code for **45S rRNA.**

3. The rRNA genes are arranged in tandem repeated clusters (i.e., the repeated genes are located next to each other).

4. **RNA polymerase I** catalyzes the formation of **45S rRNA.**

5. Another set of rRNA genes located outside of the nucleolus are transcribed by **RNA Polymerase III** to form **5S rRNA.**

B. Transfer RNA (tRNA) Genes.

1. The tRNA genes encode for **tRNAs** that are used in **protein synthesis.**

2. There are 497 tRNA genes.

3. The 497 tRNA genes are classified into 49 families based on their anticodon specificity.

C. Small Nuclear RNA (snRNA) Genes.

1. The snRNA genes encode for **snRNAs** that are components of the major GU-AG spliceosome and minor AU-AC spliceosome used in **RNA splicing during protein synthesis.**

2. The snRNAs are **uridine-rich** and are named accordingly (i.e., U1snRNA is the first snRNA to be classified).

3. There are ~70 snRNA genes that encode for **U1snRNA, U2snRNA, U4snRNA, U5snRNA, and U6 snRNA,** which are components of the **major GU-AG spliceosome.**

D. Small Nucleolar RNA (snoRNA) Genes.

1. The *snoRNA* genes encode for **snoRNAs** that direct site-specific base modifications in rRNA.

2. The **C/D box snoRNAs** direct the 2'-O-ribose methylation in rRNA.

3. The **H/ACA snoRNAs** direct site-specific pseudouridylation (uridine is isomerized to pseudouridine) of rRNA.

E. Regulatory RNA Genes.

1. The regulatory RNA genes encode for **RNAs** that are likened to mRNA because they are transcribed by RNA polymerase II, 7-methylguanosine capped, and polyadenylated.

2. The *SRA-1 (steroid receptor activator) RNA gene* encodes for **SRA-1 RNA** that functions as a co-activator of several steroid receptors.

3. The *XIST gene* encodes for **XIST RNA** that functions in X chromosome inactivation.

F. XIST Gene.

1. X chromosome inactivation is a process whereby either the **maternal X chromosome (X^M)** or **paternal X chromosome (X^P)** is inactivated, resulting in a heterochromatin structures called the **Barr body** which is located along the inside of the nuclear envelope in female cells. This inactivation process overcomes the sex difference in **X gene dosage.**

2. Males have one X chromosome and are therefore **constitutively hemizygous,** but females have two X chromosomes.

3. Gene dosage is important because many X-linked proteins interact with autosomal proteins in a variety of metabolic and developmental pathways so there needs to be a tight regulation in the amount of protein for key dosage-sensitive genes.

4. X chromosome inactivation makes females **functionally hemizygous.** X chromosome inactivation begins early in embryological development at about the **late blastula stage.** Whether the X^M or the X^P becomes inactivated is a **random and irreversible event.** However, once a progenitor cell inactivates the X^M, for example, all the daughter cells within that cell lineage will also inactivate the X^M (the same is true for the X^P). This is called **clonal selection** and means that **all females are mosaics** comprising mixtures of cells in which either the X^M or X^P is inactivated.

5. X chromosome inactivation does not inactivate all the genes; **≈20% of the total genes** on the X chromosome escape inactivation. This ≈20% of genes that remain active includes those genes that have a functional homolog on the Y chromosome (gene dosage is not affected in this case) or those genes where gene dosage is not important.

6. The mechanism of X chromosome inactivation involves two cis-acting DNA sequences called **Xic (X-inactivation center)** and **Xce (X-controlling element).**

7. The mechanism of X chromosome inactivation also involves the **XIST gene,** which encodes for **XIST RNA** that is the primary signal for spreading the inactivation along the X chromosome from which it is expressed.

G. Micro RNA (miRNA) Genes. The miRNA genes encode for **miRNAs** that block the expression of other genes.

H. Antisense RNA Genes. The antisense RNA genes encode for **antisense RNA** that binds to mRNA and physically blocks translation.

I. Riboswitch Genes. The riboswitch genes encode for **riboswitch RNA,** which binds to a target molecule, changes shape, and then switches on protein synthesis.

IV. EPIGENETIC CONTROL

A. Chemical Modification of DNA.
1. DNA can be chemically modified by **methylation of cytosine nucleotides** using **methylating enzymes.** An increased methylation of a DNA segment will make that DNA segment less likely to be transcribed into RNA and hence any genes in that DNA segment will be silenced (i.e., ↑**methylation of DNA = silenced genes**). The DNA nucleotide sequence is not altered by these modifications.
2. DNA methylation plays a crucial role in the epigenetic phenomenon called **genomic imprinting.** Genomic imprinting is the differential expression of alleles depending on whether the allele is on the paternal chromosome or the maternal chromosome.
3. We are generally accustomed to the normal situation where the allele on the paternal chromosome and the allele on the maternal chromosome are expressed or silenced at the same time. When a gene is imprinted, only the allele on the paternal chromosome is expressed while the allele on the maternal chromosome is silenced (or visa versa). Hence, there must be some mechanism that distinguishes between paternal and maternal alleles.
4. During male and female gametogenesis, male and female chromosomes must acquire some sort of **imprint** that signals the difference between paternal and maternal alleles. The role of genomic imprinting is highlighted by several rare diseases like the Prader-Willi/Angelman syndromes (see Chapter 11-II-B), complete hydatidiform moles, and Beckwith-Wiedemann syndrome that show abnormal DNA methylation patterns.

B. Clinical Considerations.
1. **Complete hydatidiform mole.**
   a. A hydatidiform mole (complete or partial) represents an abnormal placenta characterized by marked enlargement of chorionic villi. A complete mole (no embryo present) is distinguished from a partial mole (embryo present) by the amount of chorionic villous involvement.
   b. A complete mole occurs when an "empty" ovum is fertilized by a haploid sperm, which then duplicates. This results in a **46,XX karyotype** with all nuclear chromosomes of **paternal** origin. A **46,YY karyotype** complete mole does not occur since this is a genetic lethal condition.
   c. A complete mole may also occur when an "empty" ovum is fertilized by two sperm (3%–13% of complete moles). This results in a **46,XY karyotype** with all nuclear chromosomes of **paternal** origin.
   d. **Clinical features include:** gross, generalized edema of chorionic villi forming grapelike, transparent vesicles; hyperplastic proliferation of surrounding trophoblastic cells; absence of an embryo/fetus; preeclampsia during the first trimester; elevated hCG levels (>100,000 mIU/mL); and an enlarged uterus with bleeding; follow-up visits after a mole are essential because 3%–5% of moles develop into gestational trophoblastic neoplasia.

2. **Beckwith-Wiedemann syndrome (BWS).**
   a. BWS is caused by abnormal transcription and regulation of various genes located in the imprinted domain on chromosome 11p15.5.
   b. The causes of BWS involve:
      (i) The ***KCNQ1OT1* gene** on chromosome 11p15.5 which encodes for a **paternally expressed K⁺ voltage-gated ion channel**. In ≈60% of BWS cases, *KCNQ1OT1* gene hypomethylation is detectable.
      (ii) The ***H19 gene*** on chromosome 11p15.5 which encodes for a **maternally expressed H19 untranslated mRNA** that functions as a tumor suppressor. In ≈7% of BWS cases, *H19* gene hypermethylation is detectable.
      (iii) The ***CDKN1C* gene** on chromosome 11p15.5 which encodes for **cyclin-dependent kinase inhibitor 1C** that functions as a tumor suppressor. In ≈40% of familial BWS cases, *CDKN1C* gene mutations have been detected.
      (iv) In ≈20% of BWS cases, **paternal uniparental disomy** has been detected.
   c. **Prevalence.** The prevalence of BWS is 1/14,000 births.
   d. **Clinical features include:** macrosomia; macroglossia; visceromegaly; embryonal tumors (e.g., Wilms tumor, hepatoblastoma, neuroblastoma, rhabdomyosarcoma); omphalocele; neonatal hypoglycemia; ear creases/pits; adrenocortical cytomegaly; and renal abnormalities.

(C) **Chemical Modification of Histones.** Histone proteins can be chemically modified by **acetylation, methylation, phosphorylation,** or **addition of ubiquitin** (sometimes called **epigenetic marks** or **epigenetic tags**). An increased acetylation of histone proteins will make a DNA segment more likely to be transcribed into RNA and hence any genes in that DNA segment will be expressed (i.e., ↑**acetylation of histones = expressed genes**). The mechanism that determines the location and combination of epigenetic tags is unknown. This is another part of the **epigenetic code** that must be deciphered.

V. NONCODING DNA

A. **Satellite DNA.** Satellite DNA is composed of very large-sized blocks (100 kb → several Mb) of **tandemly repeated noncoding DNA. Large-scale variable number tandem repeat (VNTR) polymorphisms** are typically found in satellite DNA. The function of satellite DNA is not known.

B. **Minisatellite DNA.** Minisatellite DNA is composed of moderately-sized blocks (0.1 kb → 20 kb) of **tandemly repeated noncoding DNA. Simple VNTR polymorphisms** are typically found in minisatellite DNA. **Telomeric DNA** (a type of minisatellite DNA) allows for the replication of DNA ends in the lagging strand during chromosome replication.

C. **Microsatellite DNA (Simple Sequence Repeat; SSR).** Microsatellite DNA is composed of small-sized blocks (<100 bp) of **tandemly repeated noncoding DNA. Simple VNTR polymorphisms** are typically found in microsatellite DNA. The function of microsatellite DNA is not known.

D. **Transposons (Transposable Elements; "Jumping Genes").** Transposons are composed of **interspersed repetitive noncoding DNA,** which that make up an incredible **45% of the human nuclear genome.** Transposons are mobile DNA sequences that jump from one place in the genome to another (called **transposition**).
   1. **Types of transposons.**
      a. **Short interspersed nuclear elements (SINEs).** The **Alu repeat** (280 bp) is a SINE that is the **most abundant sequence in the human genome.** When Alu repeats are located within genes, they are confined to introns and other untranslated regions.
      b. **Long interspersed nuclear elements (LINEs).** LINE 1 (~6.1 kb) is the **most important human transposon** in that it is still actively transposing (jumping) and occasionally causes disease by disrupting important functioning genes.
      c. **Long terminal repeat (LTR) transposons.**

   d. **DNA transposons.** Most DNA transposons in humans are no longer active (i.e., they do
      not jump) and therefore are considered **transposon fossils.**

2. **Mechanism of transposition (Figure 1-2A,B).** Transposable elements jump either as double-
   stranded DNA using **conservative transposition** (a "cut-and-paste" method) or through a
   RNA intermediate using **retrotransposition.**

   a. **Conservative transposition.** In conservative transposition, the transposon jumps as dou-
      ble-stranded DNA. **Transposase** (a recombination enzyme similar to an integrase) cuts the
      transposable element at a site marked by **inverted repeat DNA sequences** (about 20 base
      pairs long). Transposase is encoded in the DNA of the transposable element. The transpo-
      son is inserted at a new location, perhaps on another chromosome. This mechanism is
      similar to the mechanism that a **DNA virus** uses in its life cycle to transform host DNA.

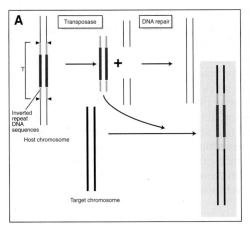

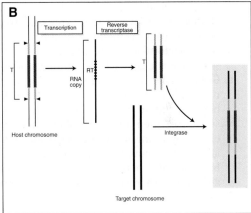

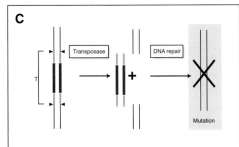

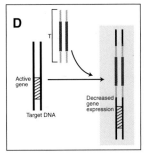

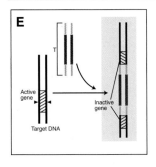

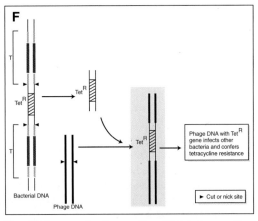

FIGURE 1-2. Transposition. (A,B) Mechanisms of transposition. (A) Conservative Transposition. **(B)** Retrotransposition.
(C-F) Transposons and genetic variability. (C) Mutation at the former site of the transposon **(D)** Level of gene expression
(E) Gene Inactivation. **(F)** Gene Transfer. T= transposon; RT: RNA code for reverse transcriptase (^): cut sites Tet^R: gene
for tetracycline resistance.

   b. Retrotransposition. In retrotransposition, the transposon jumps through a RNA intermediate. The transposon undergoes transcription, which produces a RNA copy that
   encodes a reverse transcriptase enzyme. **Reverse transcriptase** makes a double-
   stranded DNA copy of the transposon from the RNA copy. The transposon is inserted at
   a new location using the enzyme **integrase.** This mechanism is similar to the mechanism that a **RNA virus (retrovirus)** uses in its life cycle to transform host DNA.

3. Transposons and genetic variability (Figure 1-2 C-F). The main effect of transposons is to affect
   the genetic variability of the organism. Transposons can do this in several ways:
   a. Mutation at the former site of the transposon. After the transposon is cut out of its site in the
   host chromosome by transposase, the host DNA must undergo DNA repair. A mutation
   may arise at the repair site.
   b. Level of gene expression. If the transposon moves to the target DNA near an active gene,
   the transposon may affect the level of expression of that gene. While most of these
   changes in the level of gene expression would be detrimental to the organism, some of the
   changes over time might be beneficial and then spread through the population.
   c. Gene inactivation. If the transposon moves to the target DNA in the middle of a gene
   sequence, the gene will be mutated and may be inactivated.
   d. Gene transfer. If two transposons happen to be close to one another, the transposition
   mechanism may cut the ends of two different transposons. This will move the DNA
   between the two transposons to a new location. If that DNA contains a gene (or an exon
   sequence), then the gene will be transferred to a new location. This mechanism is especially important in **development of antibiotic resistance** in bacteria. Transposons in bacterial DNA can move to bacteriophage DNA, which can then spread to other bacteria. If the
   bacterial DNA between to the two transposons contains the gene for tetracycline resistance, then other bacteria will become tetracycline resistant.

Review Test

1. The human genome codes for ~30,000 genes that make up ~2% of the DNA in the human nuclear genome. The remaining nuclear genome consists of which of the following DNA elements?

(A) noncoding DNA
(B) repetitive DNA
(C) intron DNA
(D) pseudogenes
(E) satellite DNA

2. The central dogma of molecular biology is that DNA is transcribed into RNA, which is then translated into a protein. The translation takes place on the ribosomes. Which of the following RNAs are the main components of the ribosomes?

(A) tRNA
(B) snoRNA
(C) snRNA
(D) mRNA
(E) rRNA

3. A 24-year-old woman is diagnosed as having a complete molar pregnancy with enlargement of the chorionic villi and absence of an embryo. Cytogenetic analysis of the products of conception revealed a 46,XX karyotype. The molar pregnancy was caused by which one of the following?

(A) preeclampsia
(B) two haploid sets of paternal chromosomes
(C) trophoblastic neoplasia
(D) elevated hCG levels
(E) enlarged uterus

4. Which of the following is a characteristic of genomic imprinting?

(A) Most genes must bear the parent of origin imprint for proper expression.
(B) The parent of origin copy to be imprinted differs from gene to gene, and most genes require an imprint.
(C) The phenotype of a child with Prader Willi syndrome is different depending on whether the child has a deletion on chromosome 15 or UPD for the chromosome.

(D) During gamete formation, the imprint is removed from the genes and replaced with an imprint of the opposite sex.
(E) Imprinting does not disturb the primary DNA sequence.

5. Some female carriers of hemophilia B (an X-linked recessive disease) have symptoms of the disease. Which of the following is the most likely explanation for how this occurs?

(A) The X chromosome for the normal gene is inactivated in a majority of cells in the body.
(B) Triplet repeat expansion.
(C) Incomplete penetrance.
(D) Variable expressivity.

6. Heritable traits, both normal and disease producing, are determined by which of the following?

(A) introns and exons of protein-coding genes with epigenetic control
(B) RNA-coding genes under epigenetic control
(C) protein-coding genes, RNA-coding genes, and epigenetic control
(D) protein-coding genes, processed pseudogenes and retrogenes, and epigenetic control

7. The genomes of a number of organisms, including humans, have now been characterized and compared. Which of the following describes one of the findings of these endeavors?

(A) There is a correspondence between the biological complexity of an organism and the amount of noncoding DNA.
(B) There is a correspondence between the biological complexity of an organism and the amount of coding DNA.
(C) There is a correspondence between the biological complexity of an organism and the number of chromosomes.
(D) There is no correspondence between the biological complexity of an organism and the amount of coding DNA, noncoding DNA, or the number of chromosomes.

8. Genetic variability in an organism (including humans) is significantly affected by which one of the following?

(A) microsatellite DNA
(B) satellite DNA
(C) transposons
(D) heterochromatin

9. The most abundant sequence in the human genome is which one of the following?

(A) rRNA tandem repeats
(B) microsatellite DNA
(C) satellite DNA
(D) Alu repeats

10. The modification of DNA that can make transcription of a DNA segment unlikely and thus "silence" a gene containing that segment is which one of the following?

(A) methylation of cytosine nucleotides
(B) acetylation of histones
(C) retrotransposition
(D) transcription

11. Which noncoding DNA is found near the telomeres of the chromosomes?

(A) microsatellite DNA
(B) hypervariable minisatellite DNA
(C) satellite 1 DNA
(D) alpha satellite DNA

12. Which one of the following is the mechanism responsible for genomic imprinting?

(A) acetylation
(B) phosphorylation
(C) methylation
(D) transposition

Answers and Explanations

1. **The answer is (A).** Noncoding DNA such as introns, pseudogenes, and repetitive elements (such as satellite DNA) make up the rest of the genome.

2. **The answer is (E).** The main component of ribosomes is ribosomal RNA or rRNA. The other RNAs participate in the processes of transcription and translation but are not components of the ribosomes.

3. **The answer is (B).** Because of genomic imprinting, both maternal and paternal haploid sets of chromosomes are required for normal development. When there are two paternal haploid sets of chromosomes in a conceptus, a placenta will develop but not an embryo.

4. **The answer is (E).** Imprinting does not change the DNA sequence of a gene. The maternal and paternal copies of genes are mostly active or silent at the same time. The end result of a deletion of chromosome 15 or UPD is that there are no paternal copies of the gene(s) involved in the syndrome, so there is no difference in phenotype.

5. **The answer is (A).** If the normal gene is inactivated in a large enough number of cells, then there would be more defective gene products present than normal gene products and disease symptoms will be the result.

6. **The answer is (C).** Although protein-coding genes account for much of what are recognized as heritable traits, RNA-coding genes and epigenetic control are also important in gene expression, both normal and abnormal.

7. **The answer is (A).** What has been determined so far is that the amount of noncoding DNA corresponds with the biological complexity of an organism. The human genome is composed of ~98% noncoding DNA and no other organism studied to date has this amount of noncoding DNA. Humans have ~30,000 genes compared to 19,100 for the roundworm *Caenorhabditis elegans*. The number of chromosomes has no correspondence with biological complexity. For example, carp (a fish) have 100 chromosomes and humans have 46.

8. **The answer is (C).** Transposons can cause mutations in their former site when they relocate, alter gene expression at sites where they integrate, inactivate a gene by integrating somewhere in its sequence, and move pieces of nontransposon DNA to a new location in the genome. Although much of the time this is a detrimental event, sometimes the changes are beneficial and spread through the population.

9. **The answer is (D).** Alu repeats are located in the GC rich, R-band positive areas of the chromosome that contain many genes. There are many copies of the other sequences in the human genome, but they are not as abundant as the Alu sequences.

10. **The answer is (A).** Methylation of DNA is one of the primary ways that a gene can be "turned off". Methylation plays a crucial role in genomic imprinting.

11. **The answer is (B).** Hypervariable minisatellite DNA is found near the telomere and at other chromosome locations. Satellite 1 and alpha satellite DNA are found at the centromeres. Microsatellite DNA is dispersed throughout all the chromosomes.

12. **The answer is (C).** DNA is chemically modified by methylation and is less likely to be transcribed into RNA.

I. THE BIOCHEMISTRY OF NUCLEIC ACIDS (Figure 2-1)

A **nucleoside** consists of a nitrogenous base and a sugar. A **nucleotide** consists of a nitrogenous base, a sugar, and a phosphate group. DNA and RNA consist of a chain of nucleotides, which are composed of the following components:

A. Nitrogenous Bases
1. **Purines**
   a. Adenine (A)
   b. Guanine (G)
2. **Pyrimidines**
   a. Cytosine (C)
   b. Thymine (T)
   c. Uracil (U), which is found in RNA
3. **Base Pairing.** Adenine pairs with thymine or uracil (**A-T or A-U**). Cytosine pairs with guanine (**C-G**).

B. Sugars.
1. Deoxyribose, which is found in DNA
2. Ribose, which is found in RNA

C. Phosphate (PO_4^{3-}).

II. LEVELS OF DNA PACKAGING (Figure 2-2)

A. Double Helix DNA.
1. The DNA molecule is two complementary polynucleotide chains (or DNA strands) arranged as a double helix, which are held together by **hydrogen bonding** between laterally opposed base pairs (bps).
2. DNA can adopt different helical structures, which include:
   a. **A-DNA:** a right-handed helix with 11 bp/turn
   b. **B-DNA:** a right-handed helix with10 bp/turn
   c. **Z-DNA:** a left-handed helix with 12 bp/turn
3. In humans, most of the DNA is in the B-DNA form under physiological conditions.

B. Nucleosome.
1. The most fundamental unit of packaging of DNA is the nucleosome. A nucleosome consists of a histone protein octamer (two each of **H2A, H2B, H3, and H4 histone proteins**) around which 146 bp of DNA is coiled in 1.75 turns.

2. The nucleosomes are connected by spacer DNA, which results in a 10nm diameter fiber that resembles a "beads on a string" appearance by electron microscopy.
3. Histones are small proteins containing a high proportion of **lysine** and **arginine** that impart a positive charge to the proteins, which enhances its binding to negatively charged DNA. Histones bind to DNA in A-T rich regions.
4. Histone proteins have exposed N-terminal amino acid tails that are subject to modification and are crucial in regulating nucleosome structure. **Histone acetylation** of lysine by **histone acetyltransferases (HATs)** and **histone deacetylation** by **histone deacetylases (HDACs)** are the most investigated histone modifications.
5. Histone acetylation reduces the affinity between histones and DNA. An increased acetylation of histone proteins will make a DNA segment more likely to be transcribed into RNA and hence any genes in that DNA segment will be expressed (i.e., ↑**acetylation of histones = expressed genes**).

C. **30 nm Chromatin Fiber.**
   1. The 10 nm nucleosome fiber is joined by **H1 histone protein** to form a **30 nm chromatin fiber.**
   2. During interphase of mitosis, chromosomes exist as 30 nm chromatin fibers organized as **extended chromatin** (Note: when the general term "chromatin" is used, it refers specifically to the 30 nm chromatin fiber organized as extended chromatin).
   3. The extended chromatin can also form **secondary loops.**
   4. During metaphase of mitosis, chromatin undergoes **compaction.**

III. CENTROMERE

A. A centromere is a specialized nucleotide DNA sequence that binds to the mitotic spindle during cell division.

B. A major component of centromeric DNA is **α-satellite DNA,** which consists of 171 bp repeat unit.

C. A centromere is associated with a number of *cen*tromeric *p*roteins, which include: **CENP-A, CENP-B, CENP-C,** and **CENP-G.**

D. Chromosomes have a single centromere that is observed microscopically as a **primary constriction,** which is the region where sister chromatids are joined.

E. During prometaphase, a pair of protein complexes called **kinetochores** forms at the centromere and one kinetochore is attached to each sister chromatid.

F. Microtubules produced the by **centrosome** of the cell attach to the kinetochore (called **kinetochore microtubules**) and pull the two sister chromatids toward opposite poles of the mitotic cell.

IV. HETEROCHROMATIN AND EUCHROMATIN

A. **Heterochromatin** is condensed chromatin and is **transcriptionally inactive.** In electron micrographs, heterochromatin is electron dense (i.e., very black). An example of heterochromatin is the **Barr body,** which can be seen in interphase cells from females, which is the inactive X chromosome. Heterochromatin comprises ~10% of the total chromatin.
   1. **Constitutive heterochromatin** is always condensed (i.e., transcriptionally inactive) and consists of repetitive DNA found near the centromere and other regions.

   2. Facultative heterochromatin can be either condensed (i.e., transcriptionally inactive) or dispersed (i.e., transcriptionally active). An example of facultative heterochromatin is the **XY body,** which forms when both the X and Y chromosome are inactivated for ~15 days during male meiosis and the inactivated X chromosome in females.

B. Euchromatin is dispersed chromatin and comprises ~90% of the total chromatin. Of this 90%, 10% is transcriptionally active and 80% is transcriptionally inactive. When chromatin is transcriptionally active, there is weak binding to the H1 histone protein and **acetylation** of the H2A, H2B, H3, and H4 histone proteins.

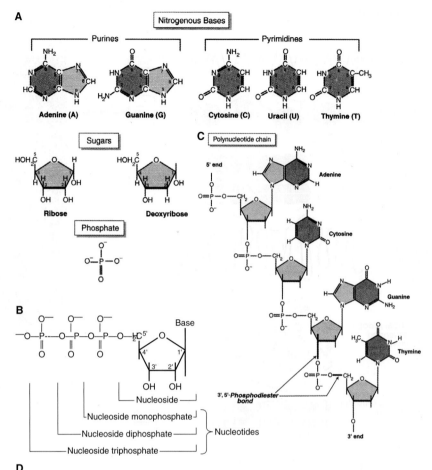

D

Base	Nucleotide (Base+sugar)	Nucleotide (Base+sugar+phosphate)			Nucleic Acid
		Monophosphate	Diphosphate	Triphosphate	
Purines					
Adenine	Adenosine	AMP	ADP	ATP	RNA
	Deoxyadenosine	dAMP	dADP	dATP	DNA
Guanine	Guanosine	GMP	GDP	GTP	RNA
	Deoxyguanosine	dGMP	dGDP	dGTP	DNA
Pyrimidines					
Cytosine	Cytidine	CMP	CDP	CTP	RNA
	Deoxycytidine	dCMP	dCDP	dCTP	DNA
Thymine	Thymidine	TMP	TDP	TTP	RNA
	Deoxythymidine	dTMP	dTDP	dTTP	DNA
Uracil	Uridine	UMP	UDP	UTP	RNA
	Deoxyuridine	dUMP	dUDP	dUTP	DNA

FIGURE 2-1. Biochemistry of DNA. (A) Structure of the biochemical components of DNA and RNA (purines, pyrimidines, sugars, and phosphate). **(B)** Diagram depicting the chemical structure of the various components of DNA. **(C)** Diagram of a DNA polynucleotide chain. The biochemical components (purines, pyrimidines, sugar, and phosphate) form a polynucleotide chain through a 3′,5′-phosphodiester bond. **(D)** Nomenclature of nucleosides and nucleotides in RNA and DNA. AMP = adenylate or adenosine 5′-monophosphate dAMP = 2′-deoxyadenosine 5′-monophosphate ADP = adenosine 5′-diphosphate dADP = 2′-deoxyadenosine 5′-diphosphate ATP = adenosine 5′-triphosphate dATP = 2′-deoxyadenosine 5′-triphosphate GMP = guanylate or guanosine 5′-monophosphate dGMP = 2′-deoxyguanosine 5′-monophosphate GDP = guanosine 5′-diphosphate dGDP = 2′-deoxyguanosine 5′-diphosphate GTP = guanosine 5′-triphosphate dGTP = 2′-deoxyguanosine 5′-triphosphate CMP = cytidylate or cytidine 5′-monophosphate dCMP = 2′-deoxycytidine 5′-monophosphate CDP = cytidine 5′-diphosphate dCDP = 2′-deoxycytidine 5′-diphosphate CTP = cytidine 5′-triphosphate dCTP = 2′-deoxycytidine 5′-triphosphate TMP = thymidylate or thymidine 5′-monophosphate dTMP = 2′-deoxythymidine 5′-monophosphate TDP = thymidine 5′-diphosphate dTDP = 2′-deoxythymidine 5′-diphosphate TTP- thymidine 5′-triphosphate dTTP = 2′-deoxythymidine 5-triphosphate UMP = uridylate or uridine 5′-monophosphate dUMP = 2′-deoxyuridine 5′-monophosphate UDP = uridine 5′-diphosphate dUDP = 2′-deoxyuridine 5′-diphosphate UTP = uridine 5′-triphosphate dUTP = 2′-deoxyuridine 5′-triphosphate.

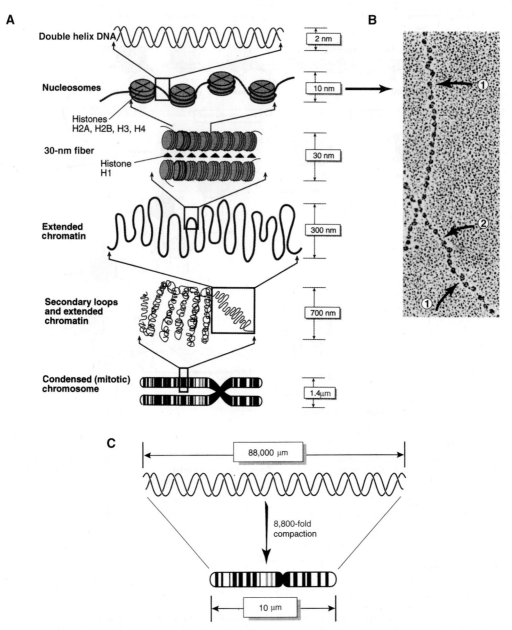

FIGURE 2-2. DNA packaging. (A) Diagram depicting the various levels of packaging of double helix DNA found within a metaphase chromosome. Double helix DNA is wound around a histone octamer of H2A, H2B, H3, and H4 to form a nucleosome. Nucleosomes are pulled together by histone H1 to form a 30 nm diameter fiber. The 30 nm fiber exists either as extended chromatin or as secondary loops within a condensed metaphase chromosome. **(B)** Electron micrograph of DNA isolated and subjected to treatments that unfold DNA to a nucleosome. The "beads on a string" appearance is the basic unit of chromatin packaging called a nucleosome. The globular structure ("bead") (arrow 1) is a histone octamer. The linear structure ("string") (arrow 2) is DNA. **(C)** Compaction of DNA in a chromosome. The double helix DNA of a chromosome is shown unraveled and stretched out measuring 88,000 um in length. During metaphase of mitosis, chromatin can become highly compacted. For example, human chromosome 1 contains about 260,000,000 bp. The distance between each base pair is 0.34 nm. So that, the physical length of the DNA comprising chromosome 1 is 88,000,000 nm or 88,000 um (260,000,000 X 0.34nm = 88,000,000 nm). During metaphase, all the chromosomes condense such that the physical length of chromosome 1 is about 10 um. Consequently, the 88,000 um of DNA comprising chromosome 1 is reduced to 10 um, resulting in an 8,800-fold compaction.

Review Test

1. In humans, the female is functionally hemizygous due to X chromosome inactivation. The inactivated chromosome is thus composed of which of the following?

(A) satellite 1 DNA
(B) beta-satellite DNA
(C) facultative heterochromatin
(D) constitutive heterochromatin
(E) euchromatin

2. The levels of DNA packaging are depicted in which of the following sequences?

(A) alpha satellite DNA, heterochromatin, centromere, euchromatin, metaphase chromosome
(B) constitutive heterochromatin, euchromatin, facultative heterochromatin, metaphase chromosome
(C) purines, pyrimidines, phosphates, nucleosome, 30 nm chromatin fiber, metaphase chromosome
(D) double helix DNA, nucleosome, 30 nm chromatin fiber, extended chromatin, metaphase chromosome
(E) double helix DNA, histones, nucleosomes, extended chromatin, metaphase chromosome

3. The nitrogenous bases that make up the nucleotides of DNA are listed in which one of the following?

(A) deoxyribose and ribose
(B) deoxyribose, ribose, and phosphate
(C) adenine, thymine, cytosine, uracil
(D) adenine, thymine, cytosine, guanine

4. Which one of the following is a major component of centromeric DNA?

(A) the Barr body
(B) the XY body
(C) alpha satellite DNA
(D) Z-DNA

5. Which one of the following is the way bases are paired in a double helix of DNA?

(A) A-T, G-C
(B) A-U, G-C
(C) A-G, C-T
(D) A-C, G-T

Answers and Explanations

1. **The answer is (C).** Both copies of the X chromosome in females are active only for a short time early in development.

2. **The answer is (E).** Double helix DNA is coiled around histones that are organized into nucleosomes, which form the extended chromatin that compacts into a metaphase chromosome.

3. **The answer is (D).** A DNA nucleotide consists of one of the nitrogenous bases adenine, thymine, cytosine or guanine, the sugar deoxyribose, and a phosphate group. Uracil and the sugar ribose are components of RNA nucleotides.

4. **The answer is (C).** The 171 base pair repeat unit of alpha satellite DNA makes up much of the centromeric DNA.

5. **The answer is (A).** In DNA, adenine (A) pairs with thymine (T) and guanine (G) pairs with cytosine (C). In RNA, adenine pairs with uracil (U).

chapter 3

Chromosome Replication

I. GENERAL FEATURES

A. Chromosome replication occurs during **S phase** of the cell cycle and involves both DNA synthesis and histone synthesis to form chromatin.

B. The timing of replication is related to the chromatin structure. An **inactive gene** packaged as **heterochromatin** is replicated **late in S phase** (e.g., in a female mammalian cell, the inactive X chromosome called the **Barr body** is packaged as heterochromatin and is replicated late in S phase).

C. An **active gene** packaged as **euchromatin** is replicated early in S phase (e.g., in the pancreatic beta cell, the insulin gene will be replicated early in S phase. However, in other cell types (e.g., hepatocytes) where the insulin gene is inactive, it will be replicated late in S phase.

D. DNA polymerases absolutely require the **3′-OH end** of a based paired primer strand as a substrate for strand extension. Therefore, a **RNA primer** (synthesized by a **primase**) is required to provide the free 3′-OH group needed to start DNA synthesis.

E. DNA polymerases copy a DNA template in the **3′ → 5′ direction,** which produces a new DNA strand in the **5′ → 3′ direction.**

F. **Deoxyribonucleoside 5′-triphosphates (dATP, dTTP, dGTP, dCTP)** pair with the corresponding base (A-T, G-C) on the template strand and form a **3′,5′-phosphodiester bond** with the **3′-OH group on the deoxyribose sugar,** which releases a **pyrophosphate.**

G. Replication is described as **semiconservative** which means that a molecule of double helix DNA contains one intact parental DNA strand and one newly synthesized DNA strand.

II. THE REPLICATION PROCESS (Figure 3-1)

A. The process starts when **topoisomerase** nicks (or breaks) a single strand of DNA, which causes DNA unwinding.

B. Chromosome replication begins at specific nucleotide sequences located throughout the chromosome called **replication origins.** Eukaryotic DNA contains **multiple replication origins** to ensure rapid DNA synthesis. Normally, the S phase of the mammalian cell cycle is **8 hours.**

C. **DNA helicase** recognizes the replication origin and opens up the double helix at that site, forming a **replication bubble** with a **replication fork** at each end. The stability of the replication fork is maintained by **single-stranded binding proteins.**

D. A replication fork contains a:
  1. **Leading strand** that is synthesized continuously by **DNA polymerase δ (delta)**.
  2. **Lagging strand** that is synthesized discontinuously by **DNA polymerase α (alpha)**. **DNA primase** synthesizes short RNA primers along the lagging strand. DNA polymerase α uses the RNA primer to synthesize DNA fragments called **Okazaki fragments.** Okazaki fragments end when they run into a downstream RNA primer. To form a continuous DNA strand from the Okazaki fragments, a **DNA repair enzyme** erases the RNA primers and replaces it with DNA. **DNA ligase** subsequently joins the all the DNA fragments together.

E. The anti-neoplastic drugs **camptothecins (e.g., irinotecan, topotecan); anthracyclines (e.g., doxorubicin); epipodophyllotoxins (e.g., etoposide VP-16, teniposide VM-26);** and **amsacrine** are topoisomerase inhibitors.

F. The anti-microbial drugs **quinolones (e.g., ciprofloxacin, ofloxacin, levofloxacin, fluoroquinolones)** are also topoisomerase inhibitors.

III. THE TELOMERE

A. The human telomere is a 3-20 kb repeating nucleotide sequence (**TTAGGG**) located at the end of a chromosome. The 3-20 kb $(TTAGGG)_n$ array is preceded by 100-300 kb of telomere—associated repeats before any unique sequence is found.

B. The telomere allows replication of linear DNA to its full length. Because DNA polymerases *cannot* synthesize in the $3' \rightarrow 5'$ direction or start synthesis *de novo*, removal of the RNA primers will always leave the 5' end of the lagging strand shorter than the leading strand. If the 5' end of the lagging strand is not lengthened, a chromosome would get progressively shorter as the cell goes through a number of cell divisions.

C. This problem of lagging strand shortening is solved by a special **RNA-directed DNA polymerase or reverse transcriptase** called **telomerase** (which has a RNA and protein component). The RNA component of telomerase carries a **CCCUAA** sequence (antisense sequence of the TTAGGG telomere) that recognizes the TTAGGG sequence on the leading strand and adds many repeats of TTAGGG to the leading strand.

D. After the repeats of TTAGGG are added to the leading strand, **DNA polymerase α** uses the TTAGGG repeats as a template to synthesize the complementary repeats on the lagging strand. Thus, the lagging strand is lengthened. **DNA ligase** joins the repeats to the lagging strand and a **nuclease** cleaves the ends to form double helix DNA with flush ends.

E. Telomerase is <u>NOT</u> utilized by a majority of **normal somatic cells,** so that chromosomes normally get successively shorter after each replication; this contributes to the finite lifespan of the cell.

F. Telomerase is utilized by **stems cells** and **neoplastic cells** so that chromosomes remain perpetually long. Telomerase may play a clinical role in **aging** and **cancer.**

IV. TYPES OF DNA DAMAGE AND DNA REPAIR

A. Chromosomal breakage refers to breaks in chromosomes due to sunlight (or ultraviolet) irradiation, ionizing irradiation, DNA cross-linking agents, or DNA damaging agents. These insults may cause **depurination of DNA, deamination of cytosine to uracil,** or **pyrimidine dimerization,** which must be repaired by DNA repair enzymes.

B. DNA repair involves **DNA excision** of the damaged site, **DNA synthesis** of the correct sequence, and **DNA ligation**. Some types of DNA repair use enzymes that do not require DNA excision.

C. The normal response to DNA damage is to stall the cell in the **G₁ phase** of the cell cycle until the damage is repaired.

D. The system that detects and signals DNA damage is a multiprotein complex called **BASC (BRCA1-associated genome surveillance complex)**. Some the components of BASC include: **ATM (ataxia telangiectasia mutated) protein, nibrin, BRCA1 protein**, and **BRCA2 protein**.

E. The clinical importance of DNA repair enzymes is illustrated by some rare inherited diseases that involve genetic defects in DNA repair enzymes such as xeroderma pigmentosa (XP), ataxia-telangiectasia, Fanconi anemia, Bloom syndrome, and hereditary nonpolyposis colorectal cancer.

F. Types of DNA damage include:
1. **Depurination.** About 5,000 purines (A's or G's) per day are lost from DNA of each human cell when the N-glycosyl bond between the purine and deoxyribose sugar-phosphate is broken. This is the most frequent type of lesion and leaves the deoxyribose sugar-phosphate with a missing purine base.
2. **Deamination of cytosine to uracil.** About 100 cytosines (C) per day are spontaneously deaminate to uracil (U). If the U is not corrected back to a C, then upon replication instead of the occurrence of a correct C-G base pairing and U-A base pairing will occur instead.
3. **Pyrimidine dimerization.** Sunlight (UV radiation) can cause covalent linkage of adjacent pyrimidines forming for example, **thymine dimers**.

V. SUMMARY TABLE OF DNA MACHINERY (Table 3-1)

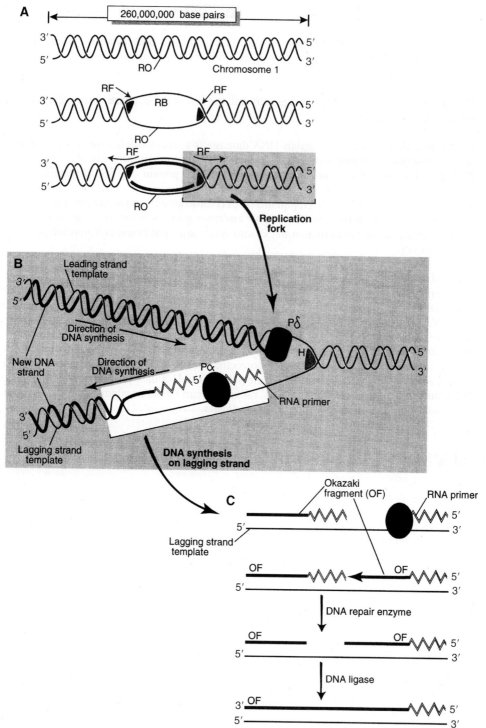

FIGURE 3-1. Replication fork. (A) A diagram of double helix DNA (Chromosome 1) at a replication origin (RO) site. DNA helicase (H) will bind at the RO and unwind the double helix into two DNA strands. This site is called a replication bubble (RB). At both ends of a replication bubble a replication fork (RF) forms. DNA synthesis occurs in a bidirectional manner from each RF (arrows). **(B)** Enlarged view of a RF at one end of the replication bubble. The leading strand serves as a template for continuous DNA synthesis in the 5′ → 3′ direction using DNA polymerase δ (Pδ). The lagging strand serves as a template for discontinuous DNA synthesis in the 5′ → 3′ direction using DNA polymerase α (Pα). Note that DNA synthesis on the leading and lagging strands is in the 5′ → 3′ direction but physically are running in opposite directions. **(C)** DNA synthesis on the lagging strand proceeds differently than on the leading strand. DNA primase synthesizes RNA primers. DNA polymerase α uses these RNA primers to synthesize DNA fragments called Okazaki fragments (OF). Okazaki fragments end when they run into a downstream RNA primer. Subsequently, DNA repair enzymes remove the RNA primers and replace it with DNA. Finally, DNA ligase joins all the Okazaki fragments together.

t a b l e   **3-1**   Summary of DNA Replication Machinery

Component	Function
Topoisomerase	Nicks (or breaks) a single strand of DNA which causes DNA unwinding
DNA helicase	Recognizes the replication fork and opens up the double helix
High Fidelity DNA-Directed DNA Polymerases	
DNA polymerase α	Synthesizes the lagging strand; $3' \rightarrow 5'$ exonuclease absent*
DNA polymerase β	Repairs DNA by base excision; $3' \rightarrow 5'$ exonuclease absent
DNA polymerase γ	Synthesizes mitochondrial DNA; $3' \rightarrow 5'$ exonuclease present
DNA polymerase δ	Synthesizes the leading strand; $3' \rightarrow 5'$ exonuclease present; repairs DNA by nucleotide and base excision
DNA polymerase ε	Repairs DNA by nucleotide and base excision; $3' \rightarrow 5'$ exonuclease present
Low Fidelity DNA-Directed DNA Polymerases	
DNA polymerase ζ	Involved in hypermutation in B and T lymphocytes
DNA polymerase η	Involved in hypermutation in B and T lymphocytes
DNA polymerase ι	Involved in hypermutation in B and T lymphocytes
DNA polymerase μ	Involved in hypermutation in B and T lymphocytes
RNA-Directed DNA Polymerase (Reverse Transcriptase)	
Telomerase	Lengthens the end of the lagging strand
LINE 1/endogenous retrovirus reverse transcriptase	Converts RNA into cDNA, which can integrate elsewhere in the genome
Primase	Synthesizes short RNA primers
Ligase	Catalyzes the formation of the $3',5'$-phosphodiester bond; joins DNA fragments
Single-stranded binding proteins	Maintain the stability of the replication fork

High fidelity = DNA sequence faithfully copied;   low fidelity = DNA sequence not faithfully copied (error prone)
* $3' \rightarrow 5'$ exonuclease = serves as proofreading activity

Review Test

1. Human cells have a finite lifespan and this contributes to the aging process. Stem cells and neoplastic cells have indefinite life spans. The reason for these observations is that chromosomes in a cell get progressively shorter with each cell division because the telomere sequences at the ends of the chromosomes get shorter with each cell division. The chromosomes in stem cells and neoplastic cells do not generally shorten with each cell division. The enzyme utilized by stem cells and neoplastic cells to lengthen the telomeres is which of the following?

(A) DNA polymerase delta
(B) DNA polymerase alpha
(C) DNA ligase
(D) topoisomerase
(E) telomerase

2. Some antineoplastic drugs act by inhibiting which of the following?

(A) DNA helicase
(B) topoisomerase
(C) telomerase
(D) DNA polymerase delta
(E) DNA polymerase alpha

3. Which one of the following is an accurate statement regarding chromosome replication?

(A) It is semiconservative.
(B) It occurs during G1 in the cell cycle.
(C) Inactive genes are replicated first.
(D) It starts with the synthesis of Okazaki fragments.

4. The leading strand of DNA in the replication fork is synthesized by which one of the following mechanisms?

(A) continuously by DNA polymerase alpha.
(B) discontinuously by DNA polymerase delta.
(C) continuously by DNA polymerase delta.
(D) discontinuously by DNA polymerase alpha.

5. The autosomal recessive disease Fanconi anemia is characterized by chromosome breakage and rearrangements and most individuals with the disease will develop some kind of cancer. Which one of the following is defective in individuals with Fanconi anemia?

(A) DNA polymerase delta
(B) DNA repair enzyme
(C) DNA ligase
(D) DNA primase

Answers and Explanations

1. **The answer is (E).** The other enzymes are involved in the replication process in general, but it is telomerase that can recognize the TTAGGG telomere sequence to that it can be replicated.

2. **The answer is (B).** Many antineoplastic drugs act by inhibiting DNA replication.

3. **The answer is (A).** Chromosome replication occurs during the S phase of the cell cycle. It starts when topoisomerase breaks a single strand of DNA, which causes DNA unwinding. Active genes are replicated early in S and inactive genes late in the S phase. Replication is semiconservative because there is one intact parental strand in a double helix of DNA and a newly synthesized strand.

4. **The answer is (C).** The lagging strand of DNA is synthesized discontinuously by DNA polymerase alpha, which synthesizes Okazaki fragments, which then have to be joined by a DNA ligase. The leading strand of DNA is synthesized continuously by DNA polymerase delta.

5. **The answer is (B).** In Fanconi anemia, DNA damage goes unrepaired and eventually reaches a point where the chromosome is unstable. The result is that there is chromosome breakage with rearrangements of chromosomal material. When a break occurs in a tumor suppressor gene, or proto-oncogenes are activated by chromosome rearrangements, the development of a malignancy is likely.

4 Mendelian Inheritance

I. AUTOSOMAL DOMINANT INHERITANCE (Figure 4-1A, B and Tables 4-1, 4-2, and 4-3)

A. Introduction. In autosomal dominant inheritance:

1. The disorder is observed in an **equal number of females and males** who are **heterozygous** for the mutant gene.
2. The characteristic **family pedigree is vertical** in that the disorder is passed from one generation to the next generation.
3. **Transmission by the mother or father** (i.e., mother-to-son; mother-to-daughter; father-to-son; father-to-daughter).
4. Although homozygotes for some autosomal dominant disorders do occur, they are rare because homozygosity for an autosomal dominant disorder is generally a genetic lethal.

B. Genetic Risk Assessment. The genetic risk associated with an autosomal dominant disorder is as follows:

1. **Example 1. Affected heterozygous mother and normal homozygous father:** In autosomal dominant disorders, the affected parent is usually a heterozygote because homozygosity for an autosomal dominant allele is frequently a genetic lethal (where those with the disorder die before they reproduce). In this example, the heterozygous mother has the disorder caused by the autosomal dominant allele "D" and the father is a normal homozygous individual. All possible combinations of alleles from the parents are shown in a Punnett square below.

	Mother	
	D	d
Father		
d	Dd	dd
d	Dd	dd

Conclusion: There is a **50% chance** (2 out of 4 children) of having a child with the autosomal dominant disorder (Dd) assuming complete penetrance. There is a **50% chance** (2 out of 4 children) of having a normal child.

2. **Example 2. Affected heterozygous mother and affected heterozygous father:** In some autosomal dominant disorders (e.g., achondroplasia), it is not unusual for individuals to choose partners who have the same condition. The parents may actually be more concerned about the chances of having a child with normal stature than one with achondroplasia. As mentioned above, homozygosity for an autosomal dominant allele is frequently a genetic lethal so that both parents with achondroplasia would be heterozygous. In this example, the heterozygous mother and the heterozygous father have the disorder caused by the

autosomal dominant allele "D." All possible combinations of alleles from the parents are shown in a Punnett square below.

	Mother	
	D	**d**
Father		
D	DD	Dd
d	Dd	dd

Conclusion: There is a **50% chance** (2 out of 4 children) of having a child with achondroplasia (Dd); a **25% chance** (1 out of 4 children) of having a normal child (dd); and a **25% chance** (1 out of 4 children) of having a child with a lethal condition (DD).

3. **Example 3. Affected homozygous mother and normal homozygous father:** In some autosomal dominant disorders (e.g., Noonan syndrome), homozygosity for an autosomal dominant allele is not a genetic lethal so that the affected individual would be homozygous. This situation is exceedingly rare and would most likely occur in cases of consanguinity, where the parents are related. In this example, the homozygous mother has the disorder caused by the autosomal dominant allele "D" and the father is a normal homozygous individual. All possible combinations of alleles from the parents are shown in a Punnett square below.

	Mother	
	D	**D**
Father		
d	Dd	Dd
d	Dd	Dd

Conclusion: There is a **100% chance** (4 out of 4 children) of having a child with Noonan syndrome (Dd).

4. If the parents of the proband are normal, the risk to the siblings of the proband is **very low** but greater than that of the general population because the possibility of germ line mosaicism exists.

C. **New Mutations.** In autosomal dominant disorders, new mutations are relatively common. In these cases, there will be an affected child with no family history of the disorder. There is a low recurrence risk (1%–2%) due to the possibility of germ line mosaicism. Germ line mosaicism is the presence of more than one cell line in the gametes in an otherwise normal parent and is the result of a mutation during the embryonic development of that parent. There is an increased risk for a new dominant mutation in fathers over 50 years of age.

D. **Reduced Penetrance.** In a reduced penetrance, many individuals have the disorder mutation but do not develop disorder symptoms. However, they can still transmit the disorder to their offspring. Example: Breast cancer whereby many women have mutations in the BRCA1 gene and BRCA2 gene but do not develop breast cancer. However, some women have mutations in the BRCA1 gene and BRCA2 gene and do develop breast cancer.

E. **Variable Expressivity.** In variable expressivity, the severity of the disorder can vary greatly between individuals. Some people may have such mild disorder that they do not know they have it until a severely affected child is born. Example: Marfan syndrome whereby a parent is tall and has long fingers, but one of his children is tall, has long fingers, and has serious cardiovascular defects.

F. **Pleiotropy.** Pleiotropy refers to a situation when a disorder has multiple effects on the body. Example: Marfan syndrome whereby the eye, skeleton, and cardiovascular system may be affected.

G. Locus heterogeneity. In locus heterogeneity, genes at more than one locus may cause the disorder. Example: Osteogenesis imperfecta whereby collagen a-1 (I) chain protein and collagen a-2(I) chain protein are encoded by the *COL1A1* gene on chromosome 17q21.3-q22 and *COL1A2* gene on chromosome 7q22.1, respectively (i.e., two separate genes located on different chromosomes). A mutation in either gene will cause osteogenesis imperfecta.

H. Example of an Autosomal Dominant Disorder. Noonan Syndrome (NS).
1. NS is an autosomal dominant genetic disorder caused by mutations in the following genes:
   a. The **PTPN11 gene** on **chromosome 12p12.1** which encodes for **tyrosine-protein phosphatase non-receptor type 11** in ≈50% of NS cases. This is an extracellular protein that plays a key role in the cellular response to growth factors, hormones, and cell adhesion molecules.
   b. The **RAF1 gene** on **chromosome 3p25** which encodes for **RAF proto-oncogene serine/threonine-protein kinase** in 3–17% of NS cases. This protein plays a key role in the signal transduction pathway for epidermal growth factor (EGF) action.
   c. The **SOS1 gene** on **chromosome 2p22 – p21,** which encodes for **son-of-sevenless homolog 1** in ≈10% of NS cases. This protein plays a key role in the signal transduction pathway for receptor tyrosine kinase action.
2. Many NS individuals have **de novo** mutations. However, an affected parent is recognized in 30% to 75% of families. In simplex cases (i.e., those with no known family history), the mutation is inherited from the father.
3. **Prevalence.** The prevalence of NS is 1/1,000 to 2,500 births.
4. **Clinical features include:** short stature, congenital heart defects, broad or webbed neck, unusual chest shape (e.g., superior pectus carinatum, inferior pectus excavatum), apparently low-set nipples, cryptorchidism in males, characteristic facial appearance (e.g., low-set, posteriorly rotated ears; vivid blue irides; widely-spaced eyes; epicanthal folds; and thick, droopy eyelids).

II. AUTOSOMAL RECESSIVE INHERITANCE (Figure 4-1C; Tables 4-1, 4-2, and 4-3)

A. Introduction. In autosomal recessive inheritance:
1. The disorder is observed in an **equal number of females and males** who are **homozygous** for the mutant gene.
2. The characteristic **family pedigree is horizontal** in that the disorder tends to be limited to a single sibship (i.e., the disorder is *not* passed from one generation to the next generation).
3. **Mother and father each transmit a recessive allele** to their sons or daughters.
4. Both parents are **obligate heterozygous carriers** whereby each parent carries one mutant allele and is asymptomatic (unless there is uniparental disomy or consanguinity, which increases the risk for autosomal recessive disorders in children).

B. Genetic Risk Assessment. The genetic risk associated with an autosomal recessive disorder is as follows:
1. **Example 1. Normal heterozygous mother and normal heterozygous father:** In autosomal recessive disorders, both parents are carriers of a single copy of the responsible gene. In this example, the mother and father are normal heterozygous carriers of the autosomal recessive allele "r." All possible combinations of alleles from the parents are shown in a Punnett square below.

	Mother	
	R	r
Father		
R	RR	Rr
r	Rr	rr

Conclusion: There is a **25% chance** (1 out of 4 children) of having a child with the autosomal recessive disorder (rr), a **50% chance** (2 out of 4 children) of having a normal child that will be a heterozygous carrier (Rr), and a **25% chance** (1 out of 4 children) of having a normal child that will *not* be a carrier (RR).

In autosomal recessive disorders, one can calculate the genetic risk for the normal children being homozygous or heterozygous. In this calculation, the child with the autosomal recessive disorder (rr = X) is eliminated from the calculation. In this example, the mother and father are normal heterozygous carriers of the autosomal recessive allele "r." All possible combinations of alleles from the parents are shown in a Punnett square below.

	Mother	
	R	r
Father		
R	RR	Rr
r	Rr	X

Conclusion: There is a **66% chance** (2 out of 3 children) of having a normal child that is a heterozygous carrier (Rr). There is a **33% chance** (1 out of 3 children) of having a normal child that is homozygous (RR).

2. **Example 2. Affected homozygous mother and normal homozygous father:** In this example, the mother has the disorder caused by the autosomal recessive allele "r" and the father is a normal homozygous individual. All the possible combinations of alleles from the parents are shown in a Punnett square below.

	Mother	
	r	r
Father		
R	Rr	Rr
R	Rr	Rr

Conclusion: There is a **100% chance** (4 out of 4 children) of having a child who is a normal heterozygous carrier (Rr).

3. **Example 3. Affected homozygous mother and normal heterozygous father:** In this example, the mother has the disorder caused by the autosomal recessive allele "r" and the father is a normal heterozygous carrier. All the possible combinations of alleles from the parents are shown in a Punnett square below.

	Mother	
	r	r
Father		
R	Rr	Rr
r	rr	rr

Conclusion: There is a **50% chance** (2 out of 4 children) of having a child with the autosomal recessive disorder (rr). There is a **50% chance** (2 out of 4 children) of having a child who is a normal heterozygous carrier (Rr).

C. Example of an Autosomal Recessive Disorder. Cystic Fibrosis (CF).

1. CF is an autosomal recessive genetic disorder caused by >1,000 mutations (almost all are point mutations or small deletions 1-84 bp) in the **CFTR gene** on **chromosome 7q31.2** for the **cystic fibrosis transmembrane conductance regulator** which functions as a chloride ion (Cl^-) channel. The Cl^- ion channel normally transports Cl^- out of the cell and H_2O follows by osmosis. The H_2O maintains the mucus in a wet and less viscous form.

2. CF is most commonly ($\approx$70% of cases in the North American population) caused by a **three base deletion** which codes for the amino acid **phenylalanine at position 508** (delta F508) such that phenylalanine is missing from CFTR. However, there are a large number

of deletions, which can cause CF, and parents of an affected child can carry different deletions of *CFTR* gene. These mutations result in absent/near absent CFTR synthesis, a block in CFTR regulation, or a destruction of Cl⁻ transport.

3. The **poly T tract/TG tract** is associated with CFTR-related disorders. The **poly T tract** is a string of thymidine bases located in intron 8 with the 5T, 7T, and 9T the most common variants. The **TG tract** is a repeat of thymidine and guanine bases just 5' of the poly T tract with repeats that commonly number 11, 12, or 13.

4. **Sweat chloride test.** The pilocarpine iontophoresis for sweat chloride is the primary diagnostic test for CF. **[Cl⁻] >60 Eq/L** on two separate occasions is diagnostic.

5. **Prevalence.** The prevalence of CF is 1/3,200 in the Caucasian population with a heterozygote carrier frequency of 1/20. CF is less common in the African American population (1/15,000) and in the Asian American population (1/31,000).

6. **Clinical features include:** production of abnormally thick mucus by epithelial cells lining the respiratory resulting in obstruction of pulmonary airways, recurrent respiratory bacterial infections, and end-stage lung disorder; pancreatic insufficiency with malabsorption; acute salt depletion, chronic metabolic alkalosis; and males are almost always sterile due to the obstruction or absence of the vas deferens.

III. X-LINKED DOMINANT INHERITANCE (Figure 4-1D and Table 4-14, 4-2, 4-3)

A. **Introduction.** In X-linked dominant inheritance:
1. The disorder is observed in **twice the number of females than males** (unless the disorder is lethal in males; then the disorder is observed only in females).
2. The characteristic **family pedigree is vertical** in that the disorder is passed from one generation to the next generation.
3. **Father-to-son transmission does *not* occur** because males have only one X chromosome (i.e., males are **hemizygous** for X-linked genes so that there is no backup copy of the gene).
4. **Males usually die (a genetic lethal).**
5. **Heterozygous females are mildly to overtly affected (never clinically normal) depending on the skew of the X chromosome inactivation.**
6. **Homozygous females (double dose) are overtly affected.**

B. **Genetic Risk Assessment.** The genetic risk associated with an X-linked dominant disorder is as follows:
1. **Example 1. Affected heterozygous mother and normal father.** In this example, the mother has the disorder (X^DX) and the father is normal (XY) because X-linked dominant disorders are usually lethal in males. All possible combinations of alleles from the parents are shown in a Punnett square below.

		Mother	
		X^D	X
Father			
	X	X^DX	XX
	Y	X^DY	XY

Conclusion: There is a **50% chance** (1 out of 2 daughters) of having a daughter with the X-linked dominant disorder. There is a **50% chance** (1 out of 2 sons) of having a son with the X-linked dominant disorder.

2. **Example 2. Normal mother and affected father.** In this example, the mother is normal (XX) and the father has the disorder (X^DY). This is a rare situation because X-linked dominant

disorders are usually lethal in males. All possible combinations of alleles from the parents are shown in a Punnett square below.

	Mother	
	X	**X**
Father		
X^D	X^{D}X	X^{D}X
Y	XY	XY

Conclusion: There is a **100% chance** (2 out of 2 daughters) of having a daughter with the X-linked dominant disorder. There is a **100% chance** (2 out of 2 sons) of having a normal son.

C. Examples of X-Linked Dominant Disorders.

1. Hypophosphatemic rickets (XLH).

a. XLH is an X-linked dominant genetic disorder caused by various mutations in the **PHEX** **gene** on **chromosome Xp22.1** for **pho**sphate regulating **e**ndopeptidase on the **X** chromosome **(PHEX)** which is a cell membrane-bound protein cleaving enzyme that degrades **phos-phatonins** (hormonelike circulating factors that increase PO$_4^{3-}$ excretion and decrease bone mineralization).

b. XLH is caused by missense, nonsense, small deletion, small insertion, or RNA splicing mutations. These mutations result in the inability of PHEX to degrade phosphatonins so that high circulating levels of phosphatonins occur, which causes **increased PO$_4^{3-}$ excretion** and **decreased bone mineralization.** These mutations also result in the underexpression of Na$^+$-PO$_4^{3-}$ Cotransporter in the kidney, which causes a **decreased PO$_4^{3-}$ absorption.**

c. Prevalence. The prevalence of XLH is 1/20,000.

d. Clinical features include: a vitamin D-resistance rickets characterized by a low serum concentration of PO$_4^{3-}$ and a high urinary concentration of PO$_4^{3-}$; short stature; dental abscesses; early tooth decay; leg deformities appeared at the time of weight-bearing; progressive departure from a normal growth rate.

2. Classic Rett syndrome (CRS).

a. CRS is an X-linked dominant genetic disorder caused by various mutations in the **MECP2 gene** on **chromosome Xq28** for **me**thyl-**Cp**G-binding protein 2 **(MECP2)** which has a methyl-binding domain (binds to 5-methylcytosine rich DNA) and a transcription repression domain (recruits other proteins that repress transcription). The MECP2 protein mediates **transcriptional repression** of various genes and **epigenetic regulation** of methylated DNA by binding to 5-methylcytosine rich DNA. Although MECP2 protein is expressed in all tissues and seems to act as a global transcriptional repressor, mutations in the MECP2 gene result in a predominately neurological phenotype.

b. CRS is caused by missense, nonsense, small deletion, and large deletion mutations. Most mutations in the *MECP2* gene occur *de novo*. These mutations result in the inability of MECP to bind 5-methylcytosine rich DNA and to repress transcription.

c. Prevalence. The prevalence of CRS in females is 1/18,000 by 15 years of age.

d. Clinical features include: a progressive neurological disorder in girls where development from birth to18 months of age is normal; later, a short period of developmental stagnation is observed followed by rapid regression in language and motor skills; purposeful use of the hands is replaced by repetitive, stereotypic hand movements (hallmark); screaming fits; inconsolable crying; autism; and paniclike attacks.

IV. X-LINKED RECESSIVE INHERITANCE (Figure 4-1E; Tables 4-1, 4-2, 4-3)

A. Introduction. In X-linked recessive inheritance:

1. The disorder is observed **only in males** (affected homozygous females are rare).

2. The characteristic **family pedigree shows skipped generations** (representing transmission through female carriers).

3. **Father-to-son transmission does *not* occur** because males have only one X chromosome (i.e., males are **hemizygous** for X-linked genes so that there is no backup copy of the gene).
4. **Males are usually sterile.**
5. **Heterozygous females are clinically normal but may be mildly affected depending on the skew of the X chromosome inactivation.**
6. **Homozygous females (double dose) are overtly affected.**

B. **Genetic Risk Assessment.** The genetic risk associated with an X-linked recessive disorder is as follows:
   1. **Example 1. Affected homozygous mother and normal father:** In this example, the mother has the disorder (X^rX^r) and the father is normal (XY). All possible combinations of alleles from the parents are shown in a Punnett square below.

	Mother	
	X^r	X^r
Father		
X	X^rX	X^rX
Y	X^rY	X^rY

   Conclusion: There is a **100% chance** (2 out of 2 daughters) of having a daughter who is a carrier of the X-linked recessive allele (X^rX). There is a **100% chance** (2 out of 2 sons) of having a son with the X-linked recessive disorder (X^rY).

   2. **Example 2. Normal heterozygous mother and normal father:** In this example, the mother is a carrier (X^rX) and the father is normal (XY). All possible combinations of alleles from the parents are shown in a Punnett square below.

	Mother	
	X^r	X
Father		
X	X^rX	XX
Y	X^rY	XY

   Conclusion: There is a **50% chance** (1 out of 2 daughters) of having a daughter who is a carrier of the X-linked recessive allele (X^rX). There is a **50% chance** (1 out of 2 sons) of having a son with the X-linked recessive disorder (X^rY).

   3. **Example 3. Normal mother and affected father:** If the father has an X-linked recessive disorder, the chances of having any children is very low because X-linked recessive males usually are sterile. However, there are a few cases of fertile X-linked recessive males. In this example, the mother is normal (XX) and the father has the disorder (X^rY). All possible combinations of alleles from the parents are shown in a Punnett square below.

	Mother	
	X	X
Father		
X^r	X^rX	X^rX
Y	XY	XY

   Conclusion: There is a **100% chance** (2 out of 2 daughters) of having a daughter who is a carrier of the X-linked recessive allele (X^rX). There is a **100% chance** (2 out of 2 sons) of having a normal son (XY) (i.e., there is no father-to-son transmission).

   4. **Example 4. Normal heterozygous mother and affected father:** In this example, the mother is a carrier (X^rX) and the father has the disorder (X^rY). This may occur in rare cases (e.g., usually consanguineous unions). All possible combinations of alleles from the parents are shown in a Punnett square below.

	Mother	
	X^r	X
Father		
X^r	X^{r}X^r	X^{r}X
Y	X^{r}Y	XY

Conclusion: There is a **50% chance** (1 out of 2 daughters) of having a daughter with the X-linked recessive disorder (X^{r}X^r); this is unusual in X-linked recessive disorders. There is a **50% chance** (1 out of 2 daughters) of having a daughter who is a carrier of the X-linked recessive allele (X^{r}X). There is a **50% chance** (1 out of 2 sons) of having a son with the X-linked recessive disorder (X^{r}Y). There is a **50% chance** (1 out of 2 sons) of having a normal son (XY).

C. **Example of X-Linked Recessive Disorder. Duchenne Muscular Dystrophy (DMD).**
1. DMD is an X-linked recessive genetic disorder caused by various mutations in the ***DMD* gene** on **chromosome Xp21.2** for **dystrophin** which anchors the cytoskeleton (actin) of skeletal muscle cells to the extracellular matrix via a transmembrane protein (**α-dystrophin and** (**β-dystrophin**) thereby stabilizing the cell membrane. The *DMD* gene is the largest known human gene.
2. DMD is caused by small deletion, large deletion, deletion of the entire gene, duplication of one of more exons, insertion, or single-based change mutations. These mutations result in absent/near absent dystrophin synthesis.
3. **Serum creatine phosphokinase (CK) measurement.** The measurement of serum CK is one of the diagnostic tests for DMD. [serum CK] $\geq$ 10 times normal is diagnostic.
4. **Skeletal muscle biopsy.** A skeletal muscle biopsy shows histological signs of fiber size variation, foci of necrosis and regeneration, hyalinization, and deposition of fat and connective tissue. Immunohistochemistry shows almost complete absence of the dystrophin protein.
5. **Prevalence.** The prevalence of DMD is 1/5,600 live male births. DMD has a 1/4,000 carrier frequency in the U.S. population, although it is difficult to calculate because ≈33% of DMD cases are new mutations.
6. **Clinical features include:** symptoms appear in early childhood with delays in sitting and standing independently; progressive muscle weakness (proximal weakness >distal weakness) often with calf hypertrophy; progressive muscle wasting; waddling gait; difficulty in climbing; wheelchair bound by 12 years of age; cardiomyopathy by 18 years of age; death by ≈30 years of age due to cardiac or respiratory failure.

V. X CHROMOSOME INACTIVATION AND X-LINKED INHERITANCE

- X chromosome inactivation is a process whereby either the **maternal X chromosome (X^M)** or **paternal X chromosome (X^P)** is inactivated resulting in a heterochromatin structures called the **Barr body** which is located along the inside of the nuclear envelope in female cells.
- This inactivation process overcomes the sex difference in **X gene dosage.** Males have one X chromosome and are therefore **constitutively hemizygous** but females have two X chromosomes.
- Gene dosage is important because many X-linked proteins interact with autosomal proteins in a variety of metabolic and developmental pathways, so there needs to be a tight regulation in the amount of protein for key dosage-sensitive genes.
- X chromosome inactivation makes females **functionally hemizygous.**
- X chromosome inactivation begins early in embryological development at about the **late blastula stage.**
- Whether the X^M or the X^P becomes inactivated is a **random and irreversible event.**

■ However, once a progenitor cell inactivates the X^M, for example, all the daughter cells within that cell lineage will also inactivate the X^M (the same is true for the X^P). This is called **clonal selection** and means that **all females are mosaics** comprising mixtures of cells in which either the X^M or X^P is inactivated.

■ X chromosome inactivation does not inactivate all the genes; **≈20% of the total genes** on the X chromosome escape inactivation. These ≈20% inactivated genes include those genes that have a functional homolog on the Y chromosome (gene dosage is not affected in this case) or those genes where gene dosage is not important.

A. **X-linked Dominant Inheritance.** In X-linked dominant inheritance, heterozygous females are mildly to overtly affected (never clinically normal).

1. **Why are heterozygous females mildly to overtly affected?** If the X chromosomes with the normal recessive gene are inactivated in a large number of cells, the female will have a large number of cells in which the one active X chromosome has the abnormal dominant gene (X^D). Therefore, the heterozygous female will be mildly to overtly affected (i.e., a range of phenotypes is possible), depending on the skew of the X chromosome inactivation.

2. **Can a female ever show overt signs of an X-linked dominant disorder?** The answer is **YES.** An X-linked dominant disorder may also be observed in females who inherit both X chromosomes with the abnormal gene (i.e., **double dose; $X^D X^D$**). In this case, the heterozygous carrier mother and the affected father pass on the X chromosome with the abnormal gene. This used to be an extremely rare event, but with the advances in treatment, more males affected with X-linked dominant disorders are surviving to reproductive age. So, the probability of inheriting an abnormal X chromosome from an affected father is increasing.

B. **X-linked Recessive Inheritance.** In X-linked recessive inheritance, heterozygous females are for the most part clinically normal.

1. **Can heterozygous females ever show signs of an X-linked recessive disorder?** The answer is **YES.** If the X chromosomes with the normal dominant gene are inactivated in a large number of cells, the female will have a large number of cells in which the one active X chromosome has the abnormal recessive gene (X^r). Therefore, the heterozygous female will be mildly affected (i.e., a range of phenotypes is possible), depending on the skew of the X chromosome inactivation.

2. **Can a female ever show overt signs of an X-linked recessive disorder?** The answer is **YES.** An X-linked recessive disorder may also be observed in females who inherit both X chromosomes with the abnormal gene (i.e., **double dose; $X^r X^r$**). In this case, the heterozygous carrier mother and the affected father pass on the X chromosome with the abnormal gene. This used to be an extremely rare event, but with the advances in treatment, more males affected with X-linked recessive disorders are surviving to reproductive age. So, the probability of inheriting an abnormal X chromosome from an affected father is increasing.

VI. THE FAMILY PEDIGREE IN VARIOUS MENDELIAN INHERITED DISORDERS (Figure 4-1)

A family pedigree is a graphic method of charting the family history using various symbols.

VII. SELECTED PHOTOGRAPHS OF MENDELIAN INHERITED DISORDERS (Figure 4-2)

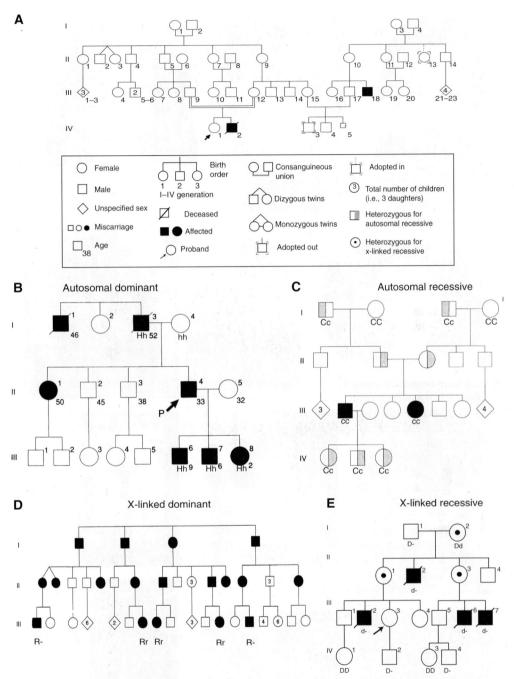

FIGURE 4-1. **(A)** A prototype family pedigree and explanation of the various symbols. **(B)** Pedigree of autosomal dominant inheritance. The disorder is observed in an **equal number of females and males** who are *heterozygous for the mutant gene.* The characteristic family pedigree is **vertical** in that the disorder is passed from one generation to the next generation. **(C)** Pedigree of autosomal recessive inheritance. The disorder is observed in an **equal number of females and males** who are *homozygous for the mutant gene.* The characteristic family pedigree is **horizontal** in that affected individuals tend to be limited to a single sibship (i.e., the disorder is not passed from one generation to the next generation). **(D)** Pedigree of X-linked dominant inheritance. The disorder is observed in **twice the number of females than males.** There is no father-to-son transmission. All daughters of an affected man will be affected because all receive the X chromosome bearing the mutant gene from their father. All sons of an affected man will be normal because they receive only the Y chromosome from the father. **(E)** Pedigree of X-linked recessive inheritance. The disorder is observed **only in males** (affected homozygous females are rare). There is no father-to-son transmission.

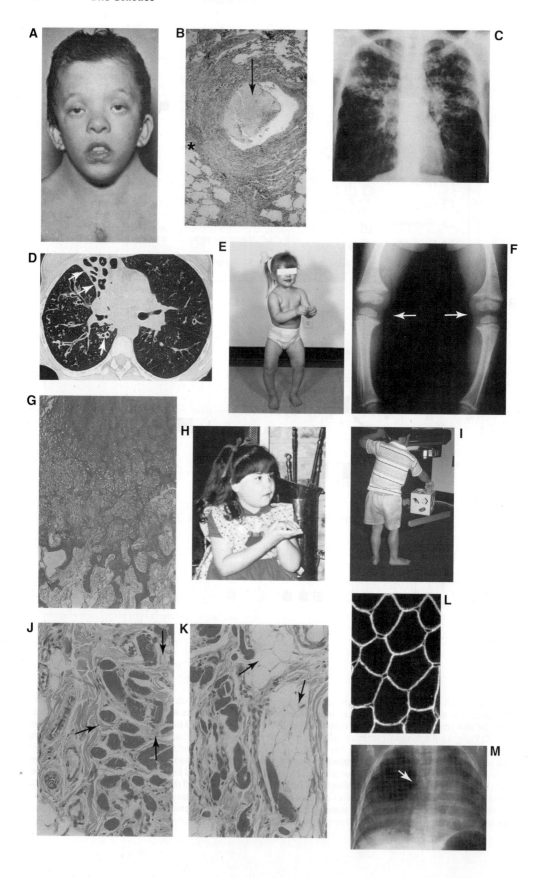

t a b l e	**4-1**	Summary Table of Major Features of Mendelian Inheritance and Mitochondrial Inheritance*	
	Sex Ratio	**Transmission Pattern**	**Other**
Autosomal dominant	Disorder is observed in an equal number of females and males	Family pedigree is vertical (disorder is passed from one generation to the next generation) Transmission by the mother or father	Homozygosity is generally a genetic lethal Nuclear inheritance
Autosomal recessive	Disorder is observed in an equal number of females and males	Family pedigree is horizontal (disorder tends to be limited to a single sibship) Mother and father each transmit a recessive allele	Both parents are obligate heterozygous carriers (unless there is uniparental disomy or consanguinity) Nuclear inheritance
X-linked dominant	Disorder is observed in twice the number of females than males (unless the disorder is lethal in males)	Family pedigree is vertical (disorder is passed from one generation to the next generation) Father-to-son transmission does not occur	Males usually die (a genetic lethal) Heterozygous females are mildly to overtly affected (never clinically normal) depending on the skew of the X chromosome inactivation Homozygous females (double dose) are overtly affected Nuclear inheritance
X-linked recessive	Disorder is observed only in males (affected homozygous females are rare)	Family pedigree shows skipped generations (representing transmission through female carriers) Father-to-son transmission does not occur	Males are usually sterile Heterozygous females are clinically normal but may be mildly affected depending on the skew of the X chromosome inactivation Homozygous females (double dose) are overtly affected Nuclear inheritance
Mitochondrial	Disorder is observed in equal number of females and males	Family pedigree is vertical (disorder is passed from one generation to the next generation) Maternal transmission only	A range of phenotypes is seen in affected females and males due to heteroplasmy Show a threshold level of mitochondria for disorder to be apparent Cells with a high requirement of ATP are more seriously affected Extranuclear inheritance

*Mitochondrial inheritance will be discussed in Chapter 6

FIGURE 4-2. Selected photographs of Mendelian inherited disorders. (A) Noonan syndrome. Photograph shows a young boy with Noonan syndrome. See text for various physical features. **(B,C,D) Cystic fibrosis. (B)** Light micrograph shows a bronchus that is filled with thick mucus and inflammatory cells (arrow). Smaller bronchi may be completely plugged by this material. In addition, surrounding the bronchus there is a heavy lymphocytic infiltration (*). **(C)** PA radiograph shows hyperinflation of both lungs, reduced size of the heart because of pulmonary compression, cyst formation, and atelectasis (collapse of alveoli) in both lungs. **(D)** CT scan shows dilated, thick-walled bronchi (large arrow), collapse of the right middle lobe (small arrows) which contains dilated airways (A). **(E,F,G) Hypophosphatemic rickets. (E)** Photograph shows a young girl with typical bowing of the legs. **(F)** Radiograph shows typical bowing of the legs, near-normal mineralization of the bones, and pronounced widening of the epiphyseal growth plates medially at the knees (arrows). **(G)** Light micrograph shows a wide epiphyseal growth plate where the chondrocytes in the zone of proliferation do not form neatly arranged stacks but instead are disorganized into irregular nests. **(H) Rett syndrome.** Photograph shows a 5-year-old girl with the typical hand position characteristic of this disorder. **(I,J,K,L,M) Duchenne muscular dystrophy. (I)** Photograph shows a young boy with pseudohypertrophy of the calves. Note how the boy braces himself by grabbing onto nearby furniture with his left hand. These patients are often late walkers. **(J)** Light micrograph shows fibrosis of the endomysium (arrows) surrounding the individual skeletal muscle cells. **(K)** Light micrograph shows the replacement of skeletal muscle cells by adipocytes (arrows) in the later stages of the disorder, which causes pseudohypertrophy. **(L)** Light micrograph (immunofluorescent staining for dystrophin) shows intense staining at the periphery skeletal muscle cells from a normal individual. In an individual with Duchenne muscular dystrophy, there would be complete absence of dystrophin staining. **(M)** Radiograph shows the typical appearance of a dilated cardiomyopathy with a water-bottle configuration and dilation of the azygous vein (arrow).

| table | 4-2 | Summary Table of Risk Assessment in Mendelian Inheritance and Mitochondrial Inheritance* |

	Parents	Children
Autosomal dominant	Affected heterozygous mother + Normal homozygous father	50% chance of having an affected child 50% chance of having a normal child
	Affected heterozygous mother + Affected heterozygous father	50% chance of having an affected child 25% chance of having a normal child 25% chance of having a lethal condition
	Affected homozygous mother + Normal homozygous father	100% chance of having an affected child
Autosomal recessive	Normal heterozygous mother + Normal heterozygous father	25% chance of having an affected child 50% chance of having a normal heterozygote child (carrier) 25% of having a normal homozygous child (noncarrier) 66% chance of having a normal heterozygote child (carrier) 33% chance of having a normal homozygous child (noncarrier)
	Affected homozygous mother + Normal homozygous father	100% chance of having a normal heterozygote child (carrier)
	Affected homozygous mother + Normal heterozygous father	50% chance of having an affected child 50% chance of having a normal heterozygote child (carrier)
X-linked dominant	Affected heterozygous mother + Normal father	50% chance of having an affected daughter 50% chance of having an affected son
	Normal mother + Affected father	100% chance of having an affected daughter 100% chance of having a normal son
X-linked recessive	Affected homozygous mother + Normal father	100% chance of having a carrier daughter 100% chance of having an affected son
	Normal heterozygous mother + Normal father	50% chance of having a carrier daughter 50% chance of having an affected son
	Normal mother + Affected father	100% chance of having a carrier daughter 100% chance of having normal son
	Normal heterozygous mother + Affected father	50% chance of having an affected daughter 50% chance of having a carrier daughter 50% chance of having an affected son 50% chance of having a normal son
Mitochondrial	Affected mother + Normal father	100% chance of having an affected daughter or son (both with a range of phenotypes)
	Normal mother + Affected father	0% chance of having an affected daughter or son

*Mitochondrial inheritance will be discussed in Chapter 6

| t a b l e | 4-3 | Partial List of Single Gene Mendelian Inherited Disorders by Type |

Autosomal Dominant	Autosomal Recessive	X-linked
Achondroplasia	α_1-Antitrypsin	**Dominant**
Acrocephalosyndactyly	Deficiency	Hypophosphatemic rickets
Adult polycystic kidney disorder	Adrenogenital	Rett syndrome
Alport syndrome	Syndromes	Goltz syndrome
Apert syndrome	Albinism	Incontinentia pigmenti
Bor syndrome	Alpha thalassemia	Orofaciodigital syndrome
Brachydactyly	Alkaptonuria	
Charcot-Marie-Tooth disorder	Argininosuccinic aciduria	**Recessive**
Cleidocranial dysplasia	Ataxia telangiectasia	Duchenne muscular
Crouzon craniofacial dysplasia	Beta thalassemia	dystrophy
Craniostenosis	Bloom syndrome	Ectodermal dysplasia
Diabetes associated with defects in	Branched chain ketonuria	Ehlers-Danlos (Type IX)
genes for glucokinase, HNF-1α, and	Childhood polycystic kidney disorder	Fabry disorder
HNF-4α	Cystic fibrosis	Fragile X syndrome
Ehlers-Danlos syndrome (Type IV)	Cystinuria	G6PD deficiency
Epidermolysis bullosa simplex	Dwarfism	Hemophilia A & B
Familial adenomatous polyposis	Ehlers-Danlos syndrome (Type VI)	Hunter syndrome
Familial hypercholesterolemia	Erythropoietic porphyria	Ichthyosis
(Type IIa)	Fanconi anemia	Kennedy syndrome
Goldenhar syndrome	Friedreich ataxia	Kinky hair syndrome
Heart-hand syndrome	Fructosuria	Lesch-Nyhan syndrome
Hereditary nonpolyposis	Galactosemia	Testicular feminization
Colorectal cancer (HNPCC)	Glycogen storage disorder	Wiskott-Aldrich syndrome
Hereditary spherocytosis	Von Gierke (Type Ia)	
Huntington disorder	Pompe (Type II)	
Marfan syndrome	Cori (Type IIIa)	
Monilethrix	Andersen (Type IV)	
Myotonic dystrophy 1 and 2	McArdle (Type V)	
Neurofibromatosis	Hers (Type VI)	
Noonan syndrome	Tarui (Type VIII)	
Osteogenesis imperfecta (Type I & IV)	Hemoglobin C disorder	
Pfeiffer syndrome	Hepatolenticular degeneration	
Piebaldism	Histidinemia	
Retinoblastoma	Homocystinuria	
Treacher Collins syndrome	Hypophosphatasia	
Spinocerebellar ataxia 1,2,3.6,7,	Hypothyroidism	
8,11,17	Junctional epidermolysis bullosa	
Uncombable hair syndrome	Juvenile myoclonus epilepsy	
Von Willebrand disorder	Lawrence Moon syndrome	
Waardenburg syndrome	Lysosomal storage disorders	
Williams-Beuren syndrome	Tay Sachs	
	Gaucher	
	Niemann-Pick	
	Krabbe	
	Sandhoff	
	Schindler	
	GM1 gangliosidosis	
	Metachromatic	
	leukodystrophy	
	Mucopolysaccharidoses	
	Hurler	
	Sanfilippo A-D	
	Morquio A&B	
	Maroteaux-Lamy	
	Sly	
	Osteogenesis imperfecta (Type II & III)	
	Oculocutaneous albinism (Type I & II)	
	Peroxisomal disorders	
	Phenylketonuria	
	Premature senility	
	Pyruvate kinase deficiency	
	Retinitis pigmentosa	
	Sickle cell anemia	
	Trichothiodystrophy	
	Tyrosinemia	
	Xeroderma pigmentosa	

1. Which of the following is the risk that an unaffected full sibling of a patient with cystic fibrosis (CF) carries a mutated CF gene?

(A) 1 in 2
(B) 1 in 4
(C) 3 in 4
(D) 2 in 3

2. What is III-1's risk to be a carrier of Alport syndrome, an X-linked recessive condition?

(A) 0
(B) 25%
(C) 50%
(D) 100%

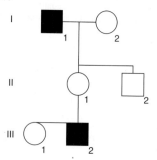

3. What is the risk that the child of a mother with cystic fibrosis will be a carrier of the disease?

(A) 100% **(C)** 50%
(B) 75%   **(D)** 25%

4. Which pedigree best represents X-linked dominant inheritance for a nonlethal condition?

(A) pedigree A
(B) pedigree B
(C) pedigree C

5. In X-linked recessive lethal disorders, the mutant gene is not always inherited from a carrier female (Haldane's rule). What approximate percentage of affected males is attributable to a new mutation?

(A) 100%
(B) 75%
(C) 66%
(D) 33%

6. Mutations in different autosomal recessively inherited genes may result in the development of leukemia in Fanconi anemia patients. Which of the following best describes why this can happen?

(A) locus heterogeneity
(B) allelic heterogeneity
(C) genotype-phenotype correlation
(D) de *novo* mutations
(E) variable expressivity

7. A 15-year-old boy is referred to a genetics clinic to rule out neurofibromatosis 1. He reports having ~25 café-au-lait spots and has started getting lumps and bumps on his skin since he hit puberty. During the family history, he describes his brother as being born with bowed legs and reports that he died at age 12 from a tumor in his neck that had been there since birth. He remembers that his brother had some birthmarks, but not nearly as many as he has. He does not recall his parents having any birthmarks, but they are not with him at the appointment.

Pedigree A

Pedigree B

Pedigree C

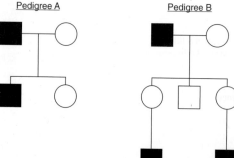

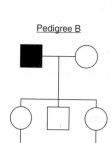

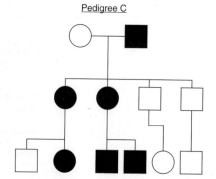

What inheritance pattern for the disease is occurring in this family?

(A) autosomal dominant
(B) autosomal recessive
(C) X-lined dominant
(D) X-linked recessive
(E) multifactorial

8. In Marfan syndrome, the affected protein, Fibrillin-1, is active in three parts of the body: the aorta, the suspensory ligaments of the lens, and the periosteum or connective tissue. This is an example of which of the following?

(A) germline mosaicism
(B) reduced penetrance
(C) variable expressivity
(D) pleiotropy
(E) locus heterogeneity

The following pedigree applies to questions 9 and 10.

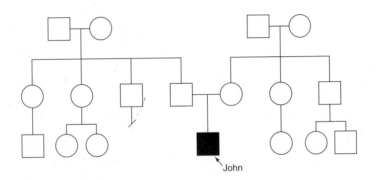

John

9. Baby John was diagnosed with achondroplasia shortly after birth. What inheritance pattern should be discussed with the parents?

(A) autosomal dominant
(B) autosomal recessive
(C) X-linked dominant
(D) X-linked recessive
(E) multifactorial

10. What is the recurrence risk for the couple to have another child with achondroplasia?

(A) 50%
(B) 25%
(C) 3%–5%
(D) 1%–2%
(E) ~0%

11. Britney and Kevin have two healthy sons, Preston and Jaden. Britney has a full brother, Brian, with G6PD deficiency. Britney's mom, Lynne, has two brothers with G6PD deficiency. Britney is currently 10 weeks pregnant by her new partner, Isaa. What is the risk the current fetus has G6PD deficiency?

(A) 1/2
(B) 1/4
(C) 1/8

(D) 1/16
(E) 1/32

12. Sally has a paternal uncle with hemophilia B, an X-linked recessive disease. Her risk of having a child with hemophilia B is best described as which of the following?

(A) near 100%
(B) near 0%
(C) 50% with all male children
(D) 50% for all children

13. Joe's brother has cystic fibrosis. What is the risk that Joe is a carrier?

(A) 1/3
(B) 2/3
(C) 1/4
(D) 1/2

14. Female carriers of X-linked recessive diseases sometimes exhibit some symptoms of the disease. The cause of this is which of the following?

(A) variable expressivity of the X-linked gene
(B) mitochondrial inheritance
(C) skewed X chromosome inactivation
(D) incomplete penetrance of the X-linked gene

Using the following information, choose the best answer to questions 15-18.

Alan has hemophilia A. His sister, Alice, has one son, Blaine. Blaine also has hemophilia A. Alan and his wife Annette have 2 children, Bart and Barbara. Barbara has a daughter, Cassie, and a son Chip. Cassie and Blaine are married and have a son, Daniel, with hemophilia A. They are now expecting fraternal twins, a boy and a girl.

15. What is Barbara's risk to be a carrier of hemophilia A?

(A) 0%
(B) 25%
(C) 50%
(D) 75%
(E) 100%

16. What is Cassie and Blaine's daughter's (the fraternal twin) risk to be affected with hemophilia A?

(A) 0%
(B) 25%
(C) 50%
(D) 75%
(E) 100%

17. What is Cassie and Blain's son's (the fraternal twin) risk to have hemophilia A?

(A) 0%
(B) 25%

(C) 50%
(D) 75%
(E) 100%

18. How are Barbara and Blaine related?

(A) first cousins
(B) first cousins once removed
(C) second cousins
(D) second cousins once removed

19. Fragile X syndrome is one of the most common causes of mental retardation in humans. It generally acts like an X-linked recessive disease, but some males do not have the disease yet they can pass it on, and some females are affected. The cause of the disease explains these observations. Fragile X syndrome is caused by which one of the following mechanisms?

(A) a deletion of the Prader-Willi/Angelman gene on the father's X chromosome
(B) a triplet repeat expansion
(C) chromosome breakage
(D) having two X chromosomes

20. In myotonic dystrophy, the severity of the disease increases with each succeeding generation. This phenomenon is called:

(A) anticipation
(B) incomplete penetrance
(C) genomic imprinting
(D) variable expressivity

Answers and Explanations

1. **The answer is (D).** If the full sibling's status was unknown, he would have a 1 in 4 risk of being unaffected and not carrying a CF mutation gene, a 2 in 4 risk of being unaffected but a carrier of a CF mutation and a 1 in 4 risk of having CF. Because he is unaffected, there are 3 possible independent outcomes. He now has a 1 in 3 chance of not carrying a mutated CF gene, but a 2 in 3 chance of being a carrier of a CF mutation.

2. **The answer is (C).** Because her mother is an obligate carrier of Alport syndrome, there is a 50% chance that she passed on the X chromosome with the mutation and a 50% chance that she passed on the normal X chromosome.

3. **The answer is (A).** The mother is homozygous for a CF mutation (aa) so she can only pass along a mutated gene (a). The father is presumably homozygous for the normal gene (AA), so he can only pass on a normal gene (A). Therefore, all their children will be heterozygotes (Aa), or carriers of a CF mutation.

4. **The answer is (C).** In Pedigree C, the condition appears in every generation in both sexes. Pedigree A is a possibility, but only males are affected and Pedigree B the condition "skips" a generation and only males are affected.

5. **The answer is (D).** In lethal disorders, all the mutated genes are lost in each generation and these represent a third of the alleles for that mutated gene. In a population at equilibrium, the number of new mutations equals the number of genes lost, so that number of new mutations replacing those lost is one-third, or 33%.

6. **The answer is (A).** Because different genes (loci) can be involved in the development of leukemia, there is locus heterogeneity.

7. **The answer is (A).** Neurofibromatosis 1 (NF1) is an autosomal dominant disease with variable expressivity. The family history and the clinical findings in the patient confirm the diagnosis of NF1. The fact that the patient's brother had it means that it probably was not due to a new mutation. One of the parents would probably be found to have some mild manifestation of the disease upon examination, as it is fully penetrant.

8. **The answer is (D).** Pleiotropy is when a gene mutation produces diverse phenotypic events. Marfan syndrome is one of the best examples of pleiotropy.

9. **The answer is (A).** Achondroplasia is an autosomal dominant disease. There is no family history because achondroplasia is often caused by a new mutation.

10. **The answer is (E).** Because the cause of achondroplasia in John is a new mutation, it is extremely unlikely to happen again so the risk is ~0.

11. **The answer is (C).** G6PD deficiency is X-linked. The risk that Britney received the mutation from her mother Lynne, an obligate carrier, is 50% or 0.5. The chance that the fetus will be a girl is 50% or ½. The chance that the girl will be a carrier is 50% if the mother is a carrier. So, $½ \times 0.5 \times ½ = ⅛$.

12. **The answer is (B).** Because Sally's father does not have hemophilia B, he does not have the X chromosome with the mutated gene to pass on to Sally. Therefore, the risk for Sally to have a child with hemophilia B is near 0.

13. **The answer is (B).** Because Joe is unaffected, he can be a carrier or not be a carrier. He has a 1 in 3 chance of not being a carrier and a 2 in 3 chance of being a carrier.

14. **The answer is (C).** If sufficient numbers of normal X chromosomes are inactivated, there may not be enough of the normal gene product present for proper functioning. In these

cases, there may be a partial or complete disease phenotype due to the fact that the majority of the gene product produced will be defective or nonfunctional.

15. **The answer is (E).** Because Barbara's daughter Cassie has a son with hemophilia A, Cassie must have received the mutation from her mother. The mutation could not have come from Blaine because he cannot pass on his X chromosome to his son. Both Barbara and Cassie are obligate carriers.

16. **The answer is (C).** Because Blaine has hemophilia A, he can only pass on an X chromosome with the mutation. Cassie is an obligate carrier with one X chromosome carrying the mutation and a normal X chromosome. The female fraternal twin can either receive a mutated X chromosome from both parents and have hemophilia A, or receive the mutated X from Blaine and a normal X from Cassie and be a carrier. The risk of being affected with hemophilia A is thus 50%.

17. **The answer is (C).** Blaine can only pass on a Y chromosome to sons. Cassie can either pass on the X chromosome with the mutation and her son will have hemophilia A, or pass on the normal X chromosome, in which case her son would be normal and not affected with hemophilia A. The risk of being affected with hemophilia A is thus 50%.

18. **The answer is (A).** Alan is Barbara's father and his sister Alice is Blaine's mother. Alan is Blaine's uncle and his daughter Barbara is Blaine's first cousin.

19. **The answer is (B).** In Fragile X syndrome the triplet repeat expansion, CGG, must reach a certain number of repeats before there is clinical manifestation of the disease. The repeat expands with succeeding generations and eventually will reach the critical number. That is why males without the disease can pass it on to subsequent generations where it appears because the threshold number of repeats has been reached. Females with a high number of repeats may also express some manifestations of the disease because of skewed X inactivation.

20. **The answer is (A).** Myotonic dystrophy is caused by a triplet repeat expansion that expands with each succeeding generation. The larger the repeat, the earlier the onset and the more severe the disease is. This phenomenon is called anticipation and differs from incomplete penetrance and variable expressivity in that once the critical repeat threshold is reached, the disease is manifested with severity depending on the number of repeats.

chapter 5

Uniparental Disomy and Repeat Mutations

I. UNIPARENTAL DISOMY (UPD, Figure 5-1)

Occurs when both copies of a chromosome are inherited from the same parent. If one copy is an identical copy of one homolog of a chromosome from a parent, then this is called **isodisomy.** If the parent passes on both homologs of a chromosome, then this is termed **heterodisomy.** In both cases, the child does not receive a copy of that chromosome from the other parent.

A. Causes.
1. UPD can be caused by the loss of one chromosome from a cell where there is a trisomy for that particular chromosome ("**trisomy rescue**").
2. UPD can also be caused when a gamete with two copies of a chromosome combines with a gamete with no copies of that chromosome.

B. Disorders. UPD can be a causative factor in a number of disorders as indicated below.
1. **Prader-Willi and Angelman (PW/A) syndromes.**
   a. Because these syndromes are mostly due to a microdeletion on chromosome 15, they are discussed in Chapter 11. However, UPD can cause the syndromes because the involved region is under control of genomic imprinting. Genomic imprinting is where the expression of a gene or genes depends on the parent of origin.
   b. If the PW/A region is deleted on the paternal chromosome 15, then Prader-Willi syndrome occurs. UPD for the maternal chromosome 15 is effectively a deletion of the paternal PW/A region so that Prader-Willi syndrome occurs.
   c. If the PW/A region is deleted on the maternal chromosome 15, then Angelman syndrome occurs. UPD for paternal chromosome 15 is effectively a deletion of the maternal PW/A region so that Angelman syndrome occurs.
2. **Beckwith-Wiedemann syndrome (BWS).** BWS can be caused by UPD where there is an excess of paternal material or loss of maternal material at chromosome 11p15. See Chapter 1-IV for more information on BWS.
3. **Autosomal recessive disorders.** In some cases, autosomal recessive disorders can be caused by UPD. Although this is a rare occurrence, UPD should be considered when only one parent is a carrier. For example, CF is an autosomal recessive disease, so both copies of the allele must have a mutation for the disease to be manifested. In UPD cases, the children have CF because they received the two mutations from the same parent.

II. UNSTABLE EXPANDING REPEAT MUTATIONS (DYNAMIC MUTATIONS)

A. Dynamic mutations are mutations that involve the **expansion of a repeat sequence** either outside or inside the gene.

B. Dynamic mutations represent a new class of mutation in humans for which there is no counterpart in other organisms. The exact mechanism by which dynamic mutations occurs is not known.

C. Although dynamic mutations may occur during mitosis resulting in mosaicism, dynamic mutations often occur only during meiosis producing the female or male gametes.

D. Threshold Length. Dynamic mutations demonstrate a **threshold length. Below a certain threshold length,** the repeat sequence is stable, does not cause disease, and is propagated to successive generations without change in length. **Above a certain threshold length,** the repeat sequence is unstable, causes disease, and is propagated to successive generations in expanding lengths.

E. Anticipation. Dynamic mutations demonstrate anticipation. Anticipation is one of the hallmarks of diseases caused by dynamic mutations. Anticipation means that a genetic disorder displays an **earlier age of onset** and/or a **greater degree of severity** in successive generations of the family pedigree.

F. Premutation Status. A normal person may have certain number of repeats that have a high likelihood of being expanded during meiosis (i.e., a permutation status) such that his offspring are at increased risk of inheriting the disease.

G. Most of dynamic mutation diseases are caused by expansion of trinucleotide repeats, although longer repeats do play a role in some diseases. Dynamic mutations are divided into two categories: highly expanded repeats outside the gene and moderately expanded CAG repeats inside the gene.

III. HIGHLY EXPANDED REPEATS OUTSIDE THE GENE

In this category of dynamic mutation, various repeat sequences (e.g., CGG, CCG, GAA, CTG, CCTG, ATTCT, or CCCCGCCCCGCG) undergo very large expansions. Below threshold length, expansions are $\approx$5 to 50 repeats. Above threshold length, expansions are $\approx$100 to 1,000 repeats. This category of dynamic mutations is characterized by the following clinical conditions.

A. Fragile X Syndrome (Martin-Bell Syndrome).
1. **Fragile X syndrome** is an X-linked recessive genetic disorder caused by a 200 to 1,000+ unstable repeat sequence of $(CGG)_n$ outside the **FMR1 gene** on **chromosome X** for the **Fragile X mental retardation 1 protein (FMRP1,** a nucleocytoplasmic shuttling protein that binds several mRNAs found abundantly in neurons.
2. The 200 to 1,000+ unstable repeat sequence of $(CGG)_n$ creates a fragile site on chromosome X, which is observed when cells are cultured in a **folate-depleted** medium. The 200 to 1,000+ unstable repeat sequence of $(CGG)_n$ has also been associated with **hypermethylation** of the *FMR1* gene so that FMRP1 is not expressed, which may lead to the phenotype of Fragile X.
3. Fragile X syndrome involves two mutation sites. **Fragile X site A** involves a 200 to 1,000+ unstable repeat sequence of $(CGG)_n$ located in a 5' UTR of the *FMR 1* **gene** on **chromosome Xq27.3. Fragile X site B** involves a 200+ unstable repeat sequence of $(CCG)_n$ located in a promoter region of the *FMR 1* **gene on chromosome Xq28.**
4. Normal *FMR1* alleles have $\approx$5 to 40 repeats. They are stably transmitted without any decrease or increase in repeat number.
5. Premutation *FMR1* alleles have $\approx$59 to 200 repeats. They are not stably transmitted. Females with permutation *FMR1* alleles are at risk for having children with Fragile X syndrome.
6. **Prevalence.** The prevalence of Fragile X syndrome is 1/4,000 males. The prevalence of Fragile X syndrome is 1/2,000 females.
7. **Clinical features include:** mental retardation (most severe in males), macroorchidism (postpubertal), speech delay, behavioral problems (e.g., hyperactivity, attention deficit), prominent

forehead and jaw, joint laxity, and large, dysmorphic ears. Fragile X syndrome is the second leading cause of inherited mental retardation (Down syndrome is the number one cause).

B. Friedreich Ataxia (FRDA).

1. FRDA is an autosomal recessive genetic disorder caused by a 600 to 1,200 unstable repeat sequence of $(GAA)_n$ in intron 1 of the **FXN gene** on **chromosome 9q13-a21.1** for the **frataxin** protein, which is located on the inner mitochondrial membrane and plays a role in the synthesis of respiratory chain complexes I through III , mitochondrial iron content, and antioxidation defense.

2. A longstanding hypothesis is that FRDA is a result of mitochondrial accumulation of iron, which may promote oxidative stress injury.

3. Normal *FXN* alleles have ≈5 to 33 repeats. They are stably transmitted without any decrease or increase in repeat number.

4. Premutation *FXN* alleles have ≈34 to 65 repeats. They are not stably transmitted. Expansion of the permutation *FXN* alleles occurs during meiosis during the production of both sperm (paternal transmission) and ova (maternal transmission) because ≈96% of FRDA individuals are homozygous for the 600 to 1,200 unstable repeat sequence of $(GAA)_n$.

5. **Prevalence.** The prevalence of FRDA is 1/50,000.

6. **Clinical features include:** degeneration of the posterior columns and spinocerebellar tracts, loss of sensory neurons in the dorsal root ganglion, slowly progressive ataxia of all four limbs with onset at 10 to 15 years of age, optic nerve atrophy, scoliosis, bladder dysfunction, swallowing dysfunction, pyramidal tract disease, cardiomyopathy (arrhythmias), and diabetes.

C. Myotonic Dystrophy Type 1 (DM1).

1. DM1 is an autosomal dominant genetic disorder caused by a >35 to 1,000 unstable repeat sequence of $(CTG)_n$ in the 3'UTR region of the **DMPK gene** on **chromosome 19q13.2-q13.3** for **myotonin-protein kinase** which is a serine-threonine protein kinase associated with intercellular conduction and impulse transmission in the heart and skeletal muscle.

2. A hypothesis is that DM1 is caused by a gain-of-function RNA mechanism in which the alternate splicing of other genes (e.g., Cl^- ion channels, insulin receptor) occurs.

3. Normal *DMPK* alleles have ≈5 to 35 repeats. They are stably transmitted without any decrease or increase in repeat number.

4. Premutation *DMPK* alleles have ≈35 to 49 repeats. They are not stably transmitted. Individuals with permutation *DMPK* alleles are at risk for having children with DM1. A child with severe DM1 (i.e., congenital DM1) most frequently inherits the expanded repeat from the mother.

5. **Prevalence.** The prevalence of DM1 is 1/100,000 in Japan, 1/10,000 in Iceland, and 1/20,000 worldwide.

6. **Clinical features include:** muscle weakness and wasting, myotonia (delayed muscle relaxation after contraction), cataracts, cardiomyopathy with conduction defects, multiple endocrinopathies, age of onset is 2 to 30 years of age, and low intelligence or dementia.

IV. MODERATELY EXPANDED CAG REPEATS INSIDE THE GENE (Figure 5-2)

In this category of dynamic mutation, a CAG repeat sequences undergoes moderate expansions. Below threshold length, expansions are ≈10 to 30 repeats. Above threshold length, expansions are ≈40 to 200 repeats. Because CAG codes for the amino acid **glutamine,** a long tract of glutamines (polyglutamine tract) will be inserted into the amino acid sequence of the protein and cause the protein to aggregate within certain cells. This category of dynamic mutations is characterized by the following clinical conditions.

A. Huntington Disease (HD).

1. HD is an autosomal dominant genetic disorder caused by a $36 \rightarrow 100+$ unstable repeat sequence of $(CAG)_n$ in the coding sequence of the *HD* **gene** on **chromosome 4p16.3** for the

Huntington protein, which is a widely expressed cytoplasmic protein present in neurons within the striatum, cerebral cortex, and cerebellum, although its precise function is unknown.

2. Because CAG codes for the amino acid glutamine, a long tract of glutamines (a polyglutamine tract) will be inserted into the Huntington protein and cause protein aggregates to form within certain cells (such as implicated in other neurodegenerative disorders).

3. Normal *HD* alleles have ≤26 repeats. They are stably transmitted without any decrease or increase in repeat number.

4. Premutation *HD* alleles have 27 to 35 repeats. They are not stably transmitted. Individuals with permutation *HD* alleles are at risk for having children with HD. A child with HD inherits the expanded repeat from the father.

5. An inverse correlation exists between the number of CAG repeats and the age of HD onset: 60 to 100 CAG repeats = juvenile onset of HD and 36 to 55 CAG repeats = adult onset of HD.

6. **Prevalence.** The prevalence of HD is 3 to 7/100,000 in populations of western European descent. HD is less common in Japan, China, Finland, and Africa.

7. **Clinical features include:** age of onset is 35 to 44 years of age, mean survival time is 15 to 18 years after onset, a movement jerkiness most apparent at movement termination, chorea (dancelike movements), memory deficits, affective disturbances, personality changes, dementia, diffuse and marked atrophy of the neostriatum due to cell death of cholinergic neurons and GABA-ergic neurons within the striatum (caudate nucleus and putamen) and a relative increase in dopaminergic neuron activity, and neuronal intranuclear aggregates. The disorder is protracted and invariably fatal. In HD, homozygotes are not more severely affected by the disorder than heterozygotes, which is an exception in autosomal dominant disorders.

B. Spinal and Bulbar Muscular Atrophy (SBMA, Kennedy Syndrome).

1. SBMA is a X-linked recessive genetic disorder caused by a >38 repeat sequence of $(CAG)_n$ in the coding sequence of the ***AR gene*** on **chromosome Xq11-q12** for the **androgen receptor** which is a member of the steroid-thyroid-retinoid superfamily of nuclear receptors and expressed in the brain, spinal cord, and muscle.

2. A hypothesis is that SBMA is caused by a gain-of-function mutation because there is a well-known syndrome called complete androgen insensitivity that is caused by a loss-of-function mutation in the AR gene.

3. Normal *AR* alleles have ≤34 repeats.

4. Premutation *AR* alleles have not been reported to date.

5. **Prevalence.** The prevalence of SBMA is <1/50,000 live males in the Caucasian and Asian population. SBMA occurs only in males.

6. **Clinical features include:** progressive loss of anterior motor neurons, proximal muscle weakness, muscle atrophy, muscle fasciculations, difficulty in swallowing and speech articulation, late-onset gynecomastia, defective spermatogenesis with reduced fertility, testicular atrophy, and a hormonal profile consistent with androgen resistance.

C. Spinocerebellar Ataxia Type 3 (SCA3, Machado-Joseph Disease).

1. SCA3 is an autosomal dominant genetic disorder caused a 52 to 86 repeat sequence of $(CAG)_n$ in coding sequence of the ***ATXN3*** gene on **chromosome 14q24.3-q31** for the **ataxin 3** protein, which is a ubiquitin-specific protease that binds and cleaves ubiquitin chains and thereby participates in protein quality control pathways in the cell.

2. A hypothesis is that SCA3 is caused by impaired protein clearance because mutant ataxin 3 forms nuclear inclusions that contain elements of the refolding and degradation machinery of the cell (i.e., chaperone and proteosome subunits).

3. Normal *ATXN3* alleles have ≤44 repeats.

4. Premutation *ATXN3* alleles have not been reported to date.

5. **Prevalence.** The prevalence of SCA3 in not known. Using a system based on genetic loci, numerous autosomal dominant ataxias have been classified (SCA1-26) and the numbers continue to grow. In general, all autosomal dominant ataxias are rare.

6. **Clinical features include:** progressive cerebellar ataxia, dysarthria, bulbar dysfunction, extrapyramidal features including rigidity and dystonia, upper and lower motor neuron signs, cognitive impairments, age of onset is 20 to 50 years of age, individuals become wheelchair bound, and nuclear inclusions are found.

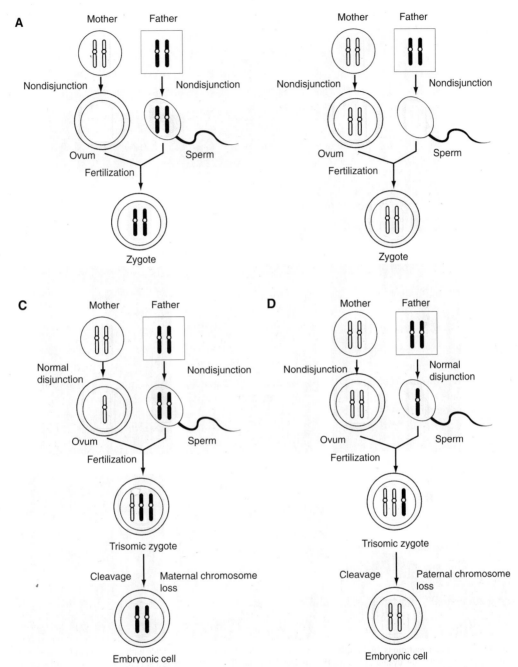

FIGURE 5-1. Uniparental disomy. (A) Maternal nondisjunction produces an ovum with no copies of a specific chromosome and paternal nondisjunction produces a sperm with two copies of the same chromosome. After fertilization, the zygote has no copies of the maternal chromosome and two copies of the paternal chromosome. **(B)** Maternal nondisjunction produces an ovum with two copies of a specific chromosome and paternal nondisjunction produces a sperm with no copies of the same chromosome. After fertilization, the zygote has two copies of the maternal chromosome and no copies of the paternal chromosome. **(C)** Maternal disjunction produces an ovum with one copy of a specific chromosome and paternal nondisjunction produces a sperm with two copies of the same chromosome. After fertilization, the zygote has three copies of the same chromosome (i.e., a trisomic zygote). The maternal chromosome is lost during mitosis of the cleavage stage, which produces embryonic cells with two copies of the paternal chromosome. **(D)** Maternal nondisjunction produces an ovum with two copies of a specific chromosome and paternal disjunction produces a sperm with one copy of the same chromosome. After fertilization, the zygote has three copies of the same chromosome (i.e., a trisomic zygote). The paternal chromosome is lost during mitosis of the cleavage stage, which produces embryonic cells with two copies of the maternal chromosome. Maternal chromosomes = white, paternal chromosomes = black.

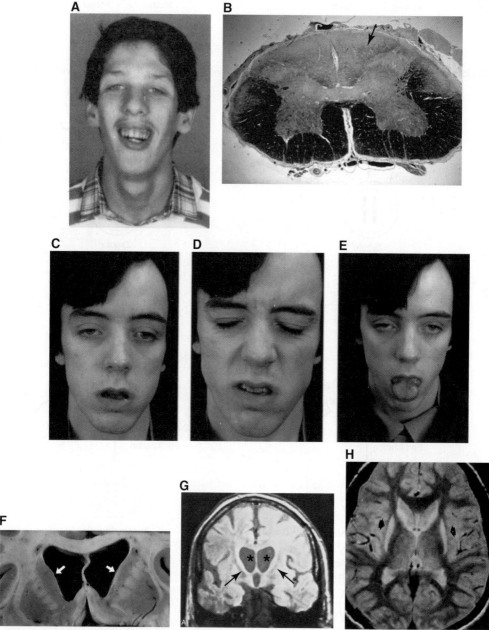

FIGURE 5-2. Dynamic mutations. (A) Fragile X syndrome. Photograph shows a young male with a prominent forehead and jaw, a long face, and large, dysmorphic ears. **(B) Friedreich ataxia.** Photograph shows a section of the spinal cord stained specifically for axons (healthy axons stain black). Note the degeneration of the posterior columns (*arrow*, whitish areas), lateral corticospinal tract, and possible the spinocerebellar tracts. **(C,D,E) Myotonic dystrophy. (C)** Photograph of a young male shows limited facial expression and inability to fully open the eyes due to weakness of the facial muscles. **(D)** Photograph of a young male shows an inability to fully close the eyes tightly or generate brow furrows due to weakness of the facial muscles. **(E)** Photograph of a young male shows fasciculations of the tongue due to delayed muscle relaxation after contraction. **(F,G,H) Huntington disorder. (F)** Pathological specimen shows atrophy of the caudate nucleus is apparent (arrows). **(G)** Coronal proton-density scan shows atrophy of the caudate nucleus (arrows) and dilatation of the frontal horns of the lateral ventricles (*). **(H)** MRI T2-weighted image shows high signal intensity in the globus pallidus (*arrows*).

Review Test

1. The repeat size of each individual's FMR-1 gene or genes is noted on the pedigree below. Which person has a full mutation in the FMR-1 gene and would have Fragile X syndrome?

(A) person A
(B) person B
(C) person C
(D) person D

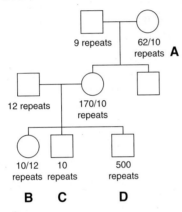

2. In myotonic dystrophy, the severity of the disease increases with each succeeding generation. Which of the following best describes this phenomenon?

(A) anticipation
(B) incomplete penetrance
(C) genomic imprinting
(D) variable expressivity

3. A conception is trisomic for chromosome 7. During the first cell division, the paternal chromosome 7 is lost, leaving two maternal chromosome 7's in the daughter cells. The daughter cells will thus have which one of the following?

(A) uniparental disomy
(B) monosomy 7
(C) double trisomy
(D) a deletion

4. One of the mechanisms by which genes are imprinted is which of the following?

(A) change in DNA sequence
(B) methylation of genes
(C) heteroplasmy
(D) trisomy rescue

5. If a sperm from the father with two copies of chromosome 15 fertilizes an egg from the mother that has no copies of chromosome 15, the conceptus would have the normal disomic complement of chromosome 15. Assuming that the conception goes to term, which one of the following phenotypes would the child have?

(A) normal
(B) Angelman syndrome
(C) Prader-Willi syndrome
(D) Beckwith-Wiedemann syndrome

6. The second leading cause of inherited mental retardation is caused by highly expanded repeats outside of a gene on the X chromosome, which may result in hypermethylation of the gene so that it is not expressed. The syndrome that results from these highly expanded repeats is which one of the following?

(A) Friedreich ataxia
(B) Myotonic dystrophy
(C) Fragile X
(D) Spinocerebellar ataxia type 3

Answers and Explanations

1. **The answer is (D).** The full mutation threshold for Fragile X is reached when there are ~200 repeats.

2. **The answer is (A).** Anticipation means that the severity of the disease increases with each generation, and/or the age of onset decreases with each generation.

3. **The answer is (A).** Because both chromosome 7's in the daughter cells came from one parent, the mother, the daughter cells have uniparental disomy.

4. **The answer is (B).** The gene from one parent that is methylated is not expressed, but the gene on the chromosome from the opposite sex is expressed.

5. **The answer is (B).** Because there are no maternal copies of the Prader-Willi/Angelman genes present, this is equivalent to a deletion of that area of the maternal chromosome 15 and Angelman syndrome will result.

6. **The answer is (C).** The CGG repeat in Fragile X is outside the FMR1 gene on the X chromosome. The threshold for the manifestation of Fragile X syndrome is ~200 repeats.

chapter 6 Mitochondrial Inheritance

I. MITOCHONDRIAL FUNCTION

Mitochondria are involved in: oxidative phosphorylation (which causes the **synthesis of adenosine triphosphate [ATP]** driven by electron transfer to oxygen), the production of acetyl coenzyme A (CoA), the tricarboxylic acid cycle, fatty acid b-oxidation, amino acid oxidation, and apoptosis. ATP is the energy source for cellular metabolism, which means that mitochondria are essential for cell functioning. There are ≈ 100 mitochondria/cell. However, different cell types have differing energy needs so that they require differing numbers of mitochondria.

A. Substrates are metabolized in the mitochondrial matrix to produce **acetyl CoA,** which is oxidized by the tricarboxylic acid cycle to carbon dioxide.

B. The energy released by this oxidation is captured by reduced nicotinamide adenine dinucleotide (NADH) and flavin adenine dinucleotide ($FADH_2$). NADH and $FADH_2$ are further oxidized, producing **hydrogen ions** and **electrons**

C. The electrons are transferred along the **electron transport chain,** which is accompanied by the outward pumping of hydrogen ions into the intermembrane space (**chemiosmotic theory).** The electron transport chain includes the following enzymes: NADH dehydrogenase (Complex I), succinate dehydrogenase (Complex II), ubiquinone-cytochrome c oxidoreductase (Complex III), and cytochrome oxidase (Complex IV).

D. The **F_0 subunit of ATP synthase** forms a transmembrane hydrogen ion pore so that hydrogen ions can flow from the intermembrane space into the matrix, where the **F_1 subunit of ATP synthase** catalyzes the reaction **$ADP + P_i \rightarrow ATP$.**

II. THE HUMAN MITOCHONDRIAL GENOME (Figure 6-1A)

A. The mitochondrial genome is completely separate from the nuclear genome. In this regard, transcription of mitochondrial DNA (mtDNA) occurs in the mitochondrial matrix, whereas transcription of nuclear DNA occurs in the nucleus.

B. The replication of mtDNA is catalyzed by DNA polymerase γ, whereas the replication of nuclear DNA is catalyzed by DNA polymerase α and λ.

C. There are several copies of the genome per mitochondrion. The human mitochondrial genome consists of mtDNA arranged as a **circular piece of double-stranded DNA (H strand and L strand)** with 16,569 base pairs and is located within the **mitochondrial matrix.**

D. In contrast to the human nuclear genome, mtDNA is not protected by histones (i.e., **histonefree**).

E. The human mitochondrial genome codes for **37 genes,** which make up **≈93% of the human mitochondrial genome.** There are **13 protein-coding genes** and **24 RNA-coding genes.** The fact that the 37 genes make up ≈93% of the human mitochondrial genome means that **≈93% of the human mitochondrial genome consists of coding DNA** and **≈7% of the human mitochondrial genome consists of noncoding DNA** (compare with the human nuclear genome in Chapter 1).

F. All human mtDNA genes **contain only exons** (i.e., no introns are present).

III. THE PROTEIN-CODING GENES

A. The protein-coding genes of the mitochondrial genome encode for 13 proteins that are not complete enzymes but are **subunits of multimeric enzyme complexes** used in electron transport and ATP synthesis. These 13 proteins synthesized on mitochondrial ribosomes.

B. The 13 proteins include the following:
   1. 7 subunits of the NADH dehydrogenase (i.e., ND1, ND2, ND3, ND4L, ND4, ND5, and ND6; Complex I)
   2. 3 subunits of the cytochrome oxidase (i.e., CO1, CO2, and CO3; Complex IV)
   3. 2 subunits of the F_0 ATP synthase (i.e., ATP synthase 6 and ATP synthase 8)
   4. 1 subunit (cytochrome b) of ubiquinone-cytochrome c oxidoreductase (i.e., CYB1; Complex III)

IV. THE RNA-CODING GENES

A. The RNA-coding genes of the mitochondrial genome encode for 24 RNAs.

B. The 24 RNAs include:
   1. 2 rRNAs (16S and 23S)
   2. 22 tRNAs (corresponding to each amino acid)

V. OTHER MITOCHONDRIAL PROTEINS

All other mitochondrial proteins (e.g., enzymes of the citric acid cycle, DNA polymerase, RNA polymerase) are **encoded by about 90 genes in the nuclear DNA, synthesized on cytoplasmic ribosomes,** and then **imported into the mitochondria.**

VI. MUTATION RATE

The mutation rate of mtDNA is about ten times that of nuclear DNA because only limited DNA repair mechanism exists in the mitochondrial matrix. This higher mutation rate may also be related to free radical production during oxidative phosphorylation.

VII. MITOCHONDRIAL INHERITANCE (Figure 6-1B; see also Chapter 4, Table 4-1)

A. **Introduction.** In mitochondrial inheritance:
   1. The disorder is observed in **both females and males who have an affected mother** (not the father).
   2. The characteristic **family pedigree is vertical** in that the disorder is passed from one generation to the next generation.
   3. There is **maternal transmission only** because sperm mitochondria do not pass into the secondary oocyte at fertilization, so that the mitochondrial genome of the zygote is determined exclusively by the mitochondria found in the cytoplasm of the unfertilized secondary oocyte.
   4. A **range of phenotypes is seen in affected females and males** because the proportion of mitochondria carrying an mtDNA mutation can differ among somatic cells (called **heteroplasmy**) and because mitochondria segregate in daughter cells independently of nuclear chromosomes.
   5. Mitochondria are the only cytoplasmic organelles in eukaryotic cells that show an inheritance independent of the nucleus (**extranuclear inheritance**). The genes located on the circular mitochondrial chromosome have an exclusively maternal transmission pattern, whereas genes located on nuclear chromosomes have Mendelian inheritance patterns.

B. **Genetic Risk Assessment (see also Chapter 4, Table 4-2).** The genetic risk associated with a mitochondrial disorder is as follows:
   1. **Example 1: Affected mother and normal father.** There is a **100% chance** of having an affected daughter with a range of phenotypes. There is a **100% chance** of having an affected son with a range of phenotypes.
   2. **Example 2: Normal mother and affected father.** There is a **0% chance** of having an affected daughter. There is a **0% chance** of having an affected son.

VIII. EXAMPLES OF MITOCHONDRIAL DISORDERS (Table 6-1)

- In general, mitochondrial disorders show a wide degree of severity among affected individuals. This variability is caused, in part, by the mixture of normal and mutant mtDNA present in a particular cell type (called **heteroplasmy**).
- When a cell undergoes mitosis, **mitochondria segregate randomly** to the daughter cells. This means that one daughter cell may receive mostly mutated mtDNA and the other daughter cell mostly normal mtDNA. The larger the population of mutant mitochondria, the more severe the disorder will be.
- Mitochondrial disorders show a **threshold level** where a critical level of mutated mitochondria must be reached before clinical symptoms appear. A woman who has not reached the threshold can still have affected children.
- Cells that have a **high requirement for ATP** (e.g., neurons and skeletal muscle) are more seriously affected by mitochondrial disorders. Mitochondrial disorders include the following:

A. **mtDNA-associated Leigh Syndrome (mtDNA-LS).**
   1. mtDNA-LS is a mitochondrial genetic disorder most commonly caused by a mutation in the **MT-ATP6** gene for the **F_0 ATP synthase 6 subunit** whereby a T → G transition occurs at **nucleotide position 8993 (T8993G),** which changes a highly conserved leucine to an arginine (L156R), or a T → C transition occurs at **nucleotide position 8993 (T8993C),** which changes a highly conserved leucine to a proline (L156P).

2. A hypothesis is that these amino acid changes in the F_o ATP synthase 6 subunit block proton (H^+ ion) translocation from the intermembrane space to the mitochondrial matrix and thereby block ATP synthesis.

3. **Prevalence.** The prevalence of mtDNA-LS is 1/140,000.

4. **Clinical features include:** hypotonia, spasticity, movement disorders, cerebellar ataxia, periphery neuropathy, signs of basal ganglia disease, hypertrophic cardiomyopathy, raised lactic acid concentration in blood and cerebrospinal fluid, death occurs by 2 to 3 years of age, and onset occurs at 3 to 12 months of age, often following a viral infection.

B. **Mitochondrial myopathy, Encephalopathy, Lactic Acidosis, and Stroke-like Episodes Syndrome (MELAS).**

1. MELAS is a mitochondrial genetic disorder caused by a mutation in the ***tRNA^Leu*** gene whereby an A → G transition occurs at **nucleotide position 3243 (A3243G).**

2. The mutated tRNA^Leu causes a reduction in the amount and the aminoacylation of the mutated tRNA^Leu, a reduction in the association of mRNA with ribosomes, and altered incorporation of leucine into mitochondrial enzymes.

3. Mitochondrial enzymes with a large number of leucine residues will have a low probability of being completely synthesized. In this regard, cytochrome oxidase (Complex IV) has been shown to be synthesized at very low rates.

4. Heteroplasmy is common and expression of the disease is highly variable.

5. **Prevalence.** The prevalence of MELAS is 1/6,250 in northern Finland.

6. **Clinical features include:** mitochondrial myopathy, encephalopathy, lactic acidosis, and strokelike episodes.

C. **Myoclonic epilepsy with ragged red fibers syndrome (MERRF).**

1. MERRF is a mitochondrial genetic disorder caused by caused by a mutation in the ***tRNA^Lys*** **gene** whereby an A → G transition occurs at **nucleotide position 8344 (A8344G).**

2. The mutated tRNA^Lys causes a **premature termination of translation** of the amino acid chain (the amount and the aminoacylation activity of the mutated tRNA^Lys is not affected).

3. Mitochondrial enzymes with a large number lysine residues will have a low probability of being completely synthesized. In this regard, **NADH dehydrogenase (Complex I)** and **cytochrome oxidase (Complex IV)** both of which have a large number of lysine residues have been shown to be synthesized at very low rates.

4. Heteroplasmy is common and expression of the disease is highly variable.

5. **Prevalence.** The prevalence of MERRF is 1/100,000 in northern Finland, 1/400,000 in northern England, and 1/400,000 in western Sweden.

6. **Clinical features include:** myoclonus (muscle twitching), seizures, cerebellar ataxia, dementia, and mitochondrial myopathy (abnormal mitochondria within skeletal muscle that impart an irregular shape and blotchy red appearance to the muscle cells, hence the term ragged red fibers).

D. **Leber's Hereditary Optic Neuropathy (LHON).**

1. LHON is a mitochondrial genetic disorder caused by 3 mtDNA missense mutations, which account for 90% of all cases worldwide and are therefore designated as **primary LHON mutations.**

2. The primary LHON mutations include the following:

    a. A mutation in the ***ND4* gene** (which encodes for subunit 4 of NADH dehydrogenase; Complex I) whereby an A → G transition occurs at **nucleotide position 11778 (A11778G).** This is the most common cause (≈50% of all LHON cases) of LHON.

    b. A mutation in the ***ND1* gene** (which encodes for subunit 1 of NADH dehydrogenase; Complex I) whereby a G → A transition occurs at **nucleotide position 3460 (G3460A).**

    c. A mutation in the ***ND 6* gene** (which encodes for subunit 6 of NADH dehydrogenase; Complex I) whereby a T → C transition occurs at **nucleotide position 14484 (T14484C).**

3. All 3 primary LHON mutations **decrease production of ATP** such that the demands of a very active neuronal metabolism cannot be met and suggest a common disease-causing mechanism.

4. Heteroplasmy is rare and expression of the disease is fairly uniform. Consequently, the family pedigree of LHON demonstrates a typical mitochondrial inheritance pattern.
5. **Prevalence.** Little data exist on the absolute prevalence of LHON. However, the prevalence of LHON is 1/33,000 in northern England.
6. **Clinical features include:** progressive optic nerve degeneration that results clinically in blindness, blurred vision, or loss of central vision; telangiectatic microangiopathy; disk pseudoedema; vascular tortuosity; onset occurs at ≈20 years of age with precipitous vision loss; and males are affected far more often than females for some unknown reason.

E. **Kearns-Sayre Syndrome (KS).**
1. KS is a mitochondrial genetic disorder caused by **partial deletions of mitochondrial DNA** (delta-mtDNA) and **duplication of mitochondrial DNA** (dup-mtDNA). The partial deletions of mtDNA have been associated with a marked reduction in the enzymatic activity of NADH dehydrogenase (Complex I), succinate dehydrogenase (Complex II), ubiquinone-cytochrome c oxidoreductase (Complex III), and cytochrome oxidase (Complex IV).
2. Heteroplasmy is common and expression of the disease is highly variable.
3. **Prevalence.** The prevalence of KS is not known, although a conservative estimate for the prevalence of all mitochondrial diseases is 1/8,500.
4. **Clinical features include:** chronic progressive external ophthalmoplegia (CPEO; degeneration of the motor nerves of the eye), pigmentary degeneration of the retina ("salt and pepper" appearance), heart block, short stature, gonadal failure, diabetes mellitus, thyroid disease, deafness, vestibular dysfunction, cerebellar ataxia, and onset occuring at ≈20 years of age.

F. **Other Common Diseases.** Mitochondrial mutations are also involved in a number of common human diseases, which include sensorineural deafness. The MELAS mutation is associated with some cases of noninsulin dependent diabetes, Alzheimer disease, Parkinson disease, and hypertrophic cardiomyopathy.

IX. SELECTED PHOTOGRAPHS OF MITOCHONDRIAL INHERITED DISORDERS (Figure 6-2)

t a b l e **6-1**	Genetic Disorders Involving Mitochondria	
Genetic Disorder	**Gene/Gene Product**	**Clinical Features**
mtDNA-associated Leigh syndrome (mtDNA-LS)	*MT-ATP6* gene/F_o ATP synthase 6 subunit	Hypotonia, spasticity, movement disorders, cerebellar ataxia, periphery neuropathy, signs of basal ganglia disease, hypertrophic cardiomyopathy, raised lactic acid concentration in blood and cerebrospinal fluid, death occurs by 2–3 years of age, and onset occurs at 3–12 months of age often following a viral infection
Mitochondrial myopathy encephalopathy lactic acidosis strokelike episodes syndrome (MELAS)	*tRNA^{Leu}* gene Cytochrome oxidase (Complex IV) is affected	Mitochondrial myopathy, encephalopathy, lactic acidosis, and strokelike episodes
Myoclonic epilepsy with ragged red fibers syndrome (MERRF)	*tRNA^{Lys}* gene NADH dehydrogenase (Complex I) and Cytochrome oxidase (Complex IV) are affected	Myoclonus (muscle twitching), seizures, cerebellar ataxia, dementia, mitochondrial myopathy (abnormal mitochondria within skeletal muscle that impart an irregular shape and blotchy red appearance to the muscle cells, hence the term ragged red fibers)
Leber's hereditary optic neuropathy (LHON)	*ND4* gene/subunit 4 of NADH dehydrogenase *ND1* gene/subunit 1 of NADH dehydrogenase *ND6* gene/subunit 6 of NADH dehydrogenase	Progressive optic nerve degeneration that results clinically in blindness, blurred vision, or loss of central vision; telangiectatic microangiopathy; disk pseudoedema; vascular tortuosity; onset occurs at ≈20 years of age with precipitous vision loss; and, affect males far more often than females or some unknown reason
Kearns-Sayre syndrome	Delta-mtDNA Dup-mtDNA NADH dehydrogenase (Complex I), Succinate dehydrogenase (Complex II), Ubiquinone-cytochrome c oxidoreductase (Complex III), and Cytochrome oxidase (Complex IV) are affected	Chronic progressive external ophthalmoplegia (CPEO; degeneration of the motor nerves of the eye), pigmentary degeneration of the retina ("salt and pepper" appearance), heart block, short stature, gonadal failure, diabetes mellitus, thyroid disease, deafness, vestibular dysfunction, cerebellar ataxia, and, onset occurs at ≈20 years of age

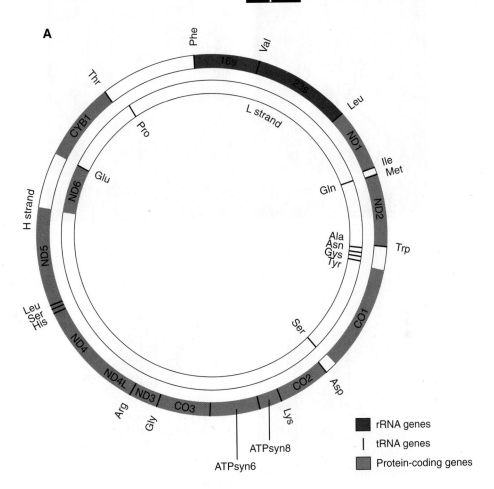

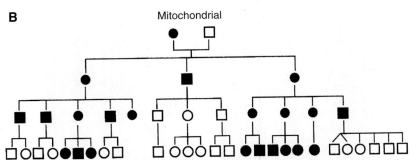

FIGURE 6-1. Mitochondrial genome and family pedigree. (A) Location of mtDNA genes and their gene products. *ND 1, ND2,ND3,ND4L,ND4, ND5, ND6* = genes for the 7 subunits of the NADH dehydrogenase complex *CO1, CO2, CO3* = genes for the 3 subunits of the cytochrome oxidase complex *ATPsyn 6, ATPsyn 8* = genes for the 2 subunits of the F₀ ATPase complex *CYB1* = gene for the 1 subunit (cytochrome b)of ubiquinone-cytochrome c oxidoreductase complex Phe = *phenylalanine tRNA gene* Val = *valine tRNA gene* Leu = *leucine tRNA gene* Ile = *isoleucine tRNA gene* Met = *methionine tRNA gene* Trp =*tryptophan tRNA gene* Asp = *asparagine tRNA gene* Lys = *lysine tRNA gene* Gly = *glycine tRNA gene* Arg = *arginine tRNA gene* His = *histidine tRNA gene* Ser = *serine tRNA gene* Thr = *threonine tRNA gene* Pro = *proline tRNA gene* Glu = glutamic acid tRNA gene* Tyr = *tyrosine tRNA gene* Cys = *cysteine tRNA gene* Asn = *asparagine tRNA gene* Ala = *alanine tRNA gene* Gln = *glutamine tRNA gene* 16S = *16S rRNA gene* 23S = *23S rRNA gene.* **(B)** Pedigree of mitochondrial inheritance. Typical pedigree seen in mitochondrial inheritance. Inheritance is matrilineal, with all children of affected mothers being affected, but not children of affected fathers. Affected fathers do not produce affected siblings.

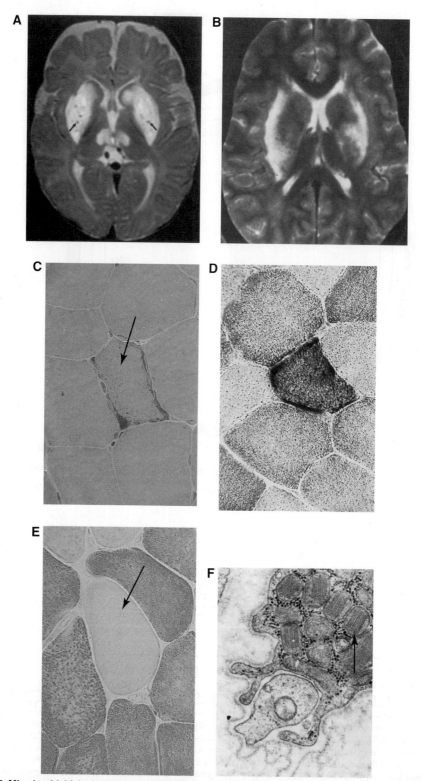

FIGURE 6-2. Mitochondrial inherited diseases. (A) mtDNA-associated Leigh syndrome. MRI (axial T2 weighted) shows bilateral hyperdensity in the caudate nucleus, putamen, and globus pallidus. Prominent perivascular spaces (arrows) can be observed scattered throughout the basal ganglia. **(B) MELAS.** MRI (axial T2 weighted) shows bilateral hyperdensity in the basal ganglia and external capsule. **(C-F) Mitochondrial myopathy. (C)** Light micrograph shows a ragged red fiber with a proliferation of subsarcolemmal mitochondria (arrow). **(D)** Light micrograph shows a ragged red fiber with normal succinate dehydrogenase staining (arrow) since succinate dehydrogenase is encoded by nuclear DNA. **(E)** Light micrograph shows a ragged red fiber with absent cytochrome oxidase staining (arrow) since cytochrome oxidase is encoded by mitochondrial DNA. **(F)** Electron micrograph shows abnormal mitochondria with paracrystalline inclusions (arrow).

Review Test

1. Which of the following describes what happens in a family where one of the parents has a mitochondrial disease?

(A) All of the children of an affected mother will inherit the disease.
(B) All of the children of an affected father will inherit the disease.
(C) Only males will be affected if the mother has the disease.
(D) Only females will be affected if the father has the disease.

2. In myoclonic epilepsy with ragged red fibers syndrome (MERRF), expression of the disease is highly variable, with some family members being severely affected while others are not affected at all. Which of the following explains the variable expression of this disease?

(A) variable expressivity
(B) incomplete penetrance
(C) heteroplasmy
(D) allelic heterogeneity

3. Jack has been diagnosed with a mitochondrial disease. His wife Sally does not have the disease and currently is pregnant. What is the risk that Jack and Sally's child will have the disease?

(A) 0%
(B) 25%
(C) 50%
(D) 100%

4. Which one of the following is a characteristic of the mitochondrial genome?

(A) It has a very low mutation rate.
(B) The DNA is bound to histones.
(C) Replication of the DNA occurs in the nucleus.
(D) There are no introns present.

5. Which of the following is a characteristic of mitochondrial disorders?

(A) They are inherited in a Mendelian fashion.
(B) There is a threshold level of mutated mitochondria that must be reached before clinical symptoms appear.
(C) All cell types in the body are equally affected by mitochondrial disorders.
(D) The degree of severity is the same for most affected individuals.

6. Which one of the following mitochondrial diseases shows the typical mitochondrial inheritance pattern?

(A) MELAS
(B) MERRF
(C) LHON
(D) KS

7. A mitochondrial disorder characterized by myoclonus, seizures, cerebellar ataxia, dementia, and mitochondrial myopathy in which the skeletal muscle cells have an irregular shape and blotchy red appearance is which one of the following?

(A) MELAS
(B) MERRF
(C) LHON
(D) KS

8. Which of the following is characteristic of the mitochondrial genome?

(A) There are 37 genes, which make up ~93% of the genome.
(B) DNA replication is the same as in the nuclear genome.
(C) The DNA in the genome is found on 23 chromosomes.
(D) Every exon has an intron.

Answers and Explanations

1. **The answer is (A).** Because all of the mitochondria in a conceptus comes from the ovum, with none coming from the sperm, all mitochondria in an individual are maternally inherited. Therefore, all of the children of a mother with a mitochondrial disease will be affected with the disease to some degree.

2. **The answer is (C).** In some mitochondrial diseases, like MERRF, there is a mixture of mitochondria, some of which carry the disease-causing mutation and some that are normal. This is called heteroplasmy. If, by chance, an individual inherits more of the normal mitochondria than abnormal, then they may exhibit a milder form of the disease or have no symptoms at all.

3. **The answer is (A).** Because mitochondrial diseases are maternally inherited, Jack cannot pass on the disease.

4. **The answer is (D).** The mitochondrial genome has a mutation rate 10 times that of the nuclear DNA, is not protected by histones, consists only of exons, and replication occurs in the mitochondrial matrix.

5. **The answer is (B).** Inheritance in mitochondrial disorders in exclusively maternal and there is a wide degree of severity among affected individuals because of heteroplasmy. Cell types that have a high ATP requirement are more seriously affected by mitochondrial disorders. If an individual does not a lot of mutated mitochondria, then the threshold where clinical symptoms appear will not be reached. When a sufficient number of mutated mitochondria are present in an individual, the threshold will be reached and the disease manifested.

6. **The answer is (C).** In Leber's hereditary optic neuropathy (LHON), heteroplasmy is rare and disease expression is relatively uniform. Heteroplasmy is common in the other mitochondrial diseases listed.

7. **The answer is (B).** In myoclonic epilepsy with ragged red fibers (MERRF) syndrome, the abnormal mitochondria in the skeletal muscle cells give them an irregular shape and a blotchy red appearance so that the muscle fibers look ragged.

8. **The answer is (A).** Mitochondrial DNA is arranged as a circular piece of double stranded DNA with no introns, only exons, with 37 genes, which make up most of the genome. DNA replication occurs in the mitochondrial matrix and is catalyzed by a different DNA polymerase than nuclear DNA.

7 Multifactorial Inherited Disorders

I. INTRODUCTION

Multifactorial inheritance involves many genes with an additive effect **(Genetic Component)** interacting with the environment **(Environmental Component)**. Both the genetic and environmental components contribute to a person inheriting the liability to develop a certain disorder. If one considers only the genetic component of a multifactorial disorder, the term **polygenic** ("many genes") is used. Inheritance patterns in multifactorial disorders usually do not conform to those seen with Mendelian inheritance. Recurrence risks are based on empiric data from population studies.

II. CLASSES OF MULTIFACTORIAL TRAITS

A. **Quantitative Traits (Figure 7-1A).** These traits are determined by many different genes along with environmental factors (e.g., diet, job exposure) that determine the phenotypic outcome. These phenotypic traits tend to follow a normal distribution in the population. Some examples of quantitative traits include: height, weight, blood pressure, and intelligence.

B. **Threshold Traits (Figure 7-1B).**
   1. These traits are relatively common isolated congenital defects with an underlying variation in liability. **Liability** is the person's predisposition for a congenital defect that may be determined by more than one gene. Liability can vary based on gender.
   2. In threshold traits, there is a **liability distribution** where a clinical effect is not seen until the threshold is reached. Individuals above this threshold have the defect because they have more of the alleles and environmental factors, which cause the defect than those below the threshold.
   3. Some examples of congenital defects, which follow this model, are: isolated cleft lip and/or cleft palate, neural tube defects (spina bifida and anencephaly), clubfoot, and pyloric stenosis. An example of an environmental component to a threshold trait is neural tube defects, where increased folic acid intake before and during pregnancy helps to prevent spina bifida and anencephaly.

C. **Common Adult Disorders.** These are disorders of adult life in which a genetic component and an environmental component play a role. Some examples include: Type 1 and Type 2 diabetes, hypertension, heart disease, cancer, alcoholism, and psychiatric disorders.

III. FACTORS AFFECTING RECURRENCE RISKS IN MULTIFACTORIAL INHERITED DISORDERS

The recurrence risk for having a multifactorial inherited disorder is the empirically derived **population risk.** However, if an individual in a family has a multifactorial inherited disorder, there are a number of factors which influence the recurrence risk in that family which include:

A. **Heritability.** Heritability (H) is the estimate of the contribution that the genetic component makes to a trait and ranges from $0 \rightarrow 1$.
   1. **If H = 0,** then the variation in a trait among individuals in a population is entirely due to the **environmental component.**
   2. **If H = 1,** then the variation in a trait among individuals in a population is entirely due to the **genetic component.**
   3. The heritability increases (i.e., H tends toward 1) as the **concordance** (two individuals having the same trait) increases in monozygotic (MZ) twins versus dizygotic (DZ) twins (↑**heritability = ↑concordance**). This is indicated in the following examples:
      a. Idiopathic seizures have a high heritability (i.e., a larger genetic component and a smaller environmental component) because the concordance for idiopathic seizures is 85% to 90% in MZ twins versus 10% to 15% in DZ twins.
      b. Cleft palate has a low heritability (i.e., smaller genetic component and a larger environmental component) because the concordance for cleft palate is 25% in MZ twins versus 10% in DZ twins.

B. **Incidence in the Population.** The incidence of a trait in a multifactorial inherited disorder often varies between populations. In Ireland, the incidence of neural tube defects is 1 in 200 births. In the United States, the incidence of neural tube defects is 1 in 1,000 births. The recurrence risk in first-degree relatives is approximately the square root of the population incidence (**recurrence risk = √ population incidence).**

C. **Sex Bias.** Multifactorial inherited disorders are often more common in one sex than in the other (i.e., sex biased). Examples of male sex biased disorders/malformations include: pyloric stenosis, cleft lip with or without cleft palate, spina bifida, and anencephaly. Examples of female sex biased disorders/malformations include: congenital hip dysplasia and idiopathic scoliosis. In sex-biased disorders, the recurrence risk is higher in the following conditions:
   1. If an affected child in the family is of the more-affected sex, then the recurrence risk to a same-sexed sibling is higher than the recurrence risk to another-sex sibling. The recurrence risk to either-sex sibling is greater than the general population risk for the condition.
   2. If an affected child in the family is of the less-affected sex, then the recurrence risk to another-sexed sibling is higher than the recurrence risk to a same-sexed sibling. The recurrence risk to either-sex sibling is greater than the general population risk for the condition.

Example 7-1. Male-Biased Pyloric Stenosis
Pyloric stenosis is an obstruction of the area between the stomach and the intestines that causes severe feeding problems and is more common in males (male-biased).

   1. If a male child in a family was born with pyloric stenosis, then the recurrence risk to his future brother is 3.8%, whereas the recurrence risk to his future sister is 2.7%.

   However,

   2. If a female child in a family was born with pyloric stenosis, then the recurrence risk to her future brother is 9.2%, whereas the recurrence risk to her future sister is 3.8%.

Example 7-2. Female-Biased Idiopathic Scoliosis

Idiopathic scoliosis is a lateral deviation of the vertebral column that involves both deviations and rotation of the vertebral bodies due to unknown causes.

1. If a female child in a family was born with idiopathic scoliosis, then the recurrence risk to her future sister is 5.3%, whereas the recurrence risk to her future brother is 1.1%

   However,

2. If a male child in a family was born with idiopathic scoliosis, then the recurrence risk to his future sister is 7.5%, whereas the recurrence risk to his future brother is 5.4%.

D. **Degree of Relationship.** The recurrence risk in a multifactorial inherited disorder increases as the closeness of the relationship to the proband increases (↑**recurrence risk as ↑closeness of relationship**). For example, in cleft palate the recurrence risk is:
   1. 4% for first-degree relatives
   2. 0.7% for second-degree relatives
   3. 0.3% for third-degree relatives which ≈ equals the population incidence.

E. **Number of Affected Relatives.** The recurrence risk in a multifactorial inherited disorder increases as the number of affected relatives increases (↑**recurrence risk as ↑number of affected relatives**). This is because multiple affected relatives suggest a high liability in the family for a multifactorial trait. For example, in cleft lip and cleft palate the recurrence risk is:
   1. 4% after having one affected child
   2. 10% after having two affected children

F. **Severity of the Disorder.** The recurrence risk in a multifactorial inherited disorder increases as the severity of the disorder increases (↑**recurrence risk as ↑severity of disorder**). This is because a more severely affected individual suggests a high liability for the trait. For example, in bilateral cleft lip and cleft palate the recurrence risk is:
   1. 2.5% after having a child with unilateral cleft lip
   2. 5.6% after having a child with bilateral cleft lip and cleft palate.

G. **Consanguinity (kinship).** The recurrence risk in a multifactorial inherited disorder increases as the consanguinity of the mating partners increases (↑**recurrence risk as ↑consanguinity**). This is because close consanguinity mating partners have high likelihood of sharing predisposing genes (i.e., a genetic relationship). For example, if the mating partners are first cousins the risk of having a child with a birth defect is 6% to 10%. Whereas, the general population risk of having a child with a birth defect is 3% to 5%.

IV. SOME COMMON MULTIFACTORIAL CONDITIONS

A. **Type 1 Diabetes.**
   1. The characteristic dysfunction in this disorder is an **autoimmune destruction of pancreatic beta cells** that produce insulin. This results clinically in hyperglycemia, ketoacidosis, and exogenous insulin dependence.
   2. Type 1 diabetes demonstrates an association with the highly polymorphic **HLA (human leukocyte antigen) class II genes,** which play a role in immune responsiveness and account for ≈40% of familial clustering of Type 1diabetes. The specific loci involved in Type 1 diabetes are called **HLA-DR3** and **HLA-DR4 loci** located on **chromosome 6p.** HLA-DR3 and HLA-DR4 loci code for **MHC Class II cell-surface glycoproteins** that are expressed on antigen-presenting cells (e.g., macrophages). >90% of Type 1 diabetic patients carry either the **HLA DR3-DQ2 allele** or the **HLA DR4-DQ8 allele.**
   3. It is *hypothesized* that alleles closely linked to *HLA-DR3* and *HLA-DR4* loci somehow alter the immune response such that the individual has an immune response to an environmental antigen (e.g., virus). The immune response "spills over" and leads to the destruction

of pancreatic beta cells. Markers for immune destruction of pancreatic beta cells include **autoantibodies to glutamic acid decarboxylase (GAD$_{65}$), insulin,** and **tyrosine phosphatases IA-2 and IA-2β.** At present, it is not known whether the autoantibodies play a causative role in the destruction of the pancreatic beta cell, or whether the autoantibodies form secondarily after the pancreatic beta cells have been destroyed.

4. Clinical features include: neuropathy, retinopathy leading to blindness, and nephropathy leading to kidney failure.

B. Type 2 Diabetes.

1. Type 2 diabetes accounts for ≈90% of all cases of diabetes. In contrast to Type 1 diabetes, there is almost always some insulin production. Type 2 diabetics develop insulin resistance, a condition where the cells have reduced ability to use insulin.

2. The disorder typically occurs in adults over 40 with the greatest risk factors being obesity and a family history. Type 2 diabetes can usually be treated effectively by a combination of diet modification and exercise. A regular exercise regimen can substantially reduce the risk of developing the disorder, even in those with a family history. This is because regular exercise helps prevent weight gain but it also aids in preventing the development of insulin resistance.

3. There is a strong genetic component to Type 2 diabetes because the concordance rate in MZ twins is ≈90%. The recurrence risk for first-degree relatives is high (15% → 40%). Populations that have adopted Western diet and activity patterns show an increased incidence of the disorder, most likely due to an increase in obesity.

4. A number of genes have been linked to Type 2 diabetes but the genetic component of the disorder is not completely understood.

C. Hypertension. Hypertension is a major factor in cardiovascular disorder and strokes. The heritability of hypertension is ≈20% to 40%. Although a number of genes have been linked to hypertension, hypertension involves complex physiological processes that involve many genes. The role of environmental factors is also recognized in the etiology of hypertension (e.g., sodium in the diet, physical activity, and weight gain).

D. Heart Disease. Some genes known to play a role in heart disease are involved with the regulation of lipoproteins in the circulation. There are a number of risk factors for heart disease (e.g., cigarette smoking, obesity, hypertension, high cholesterol, and a positive family history). The risk of developing heart disease increases in an individual who has:

1. A first-degree affected male relative (recurrence risk = ≈2× the general population risk)

2. A first-degree affected female (least affected sex) relative (recurrence risk ≥2× the general population risk)

3. Many affected relatives (recurrence risk ≥2× the general population risk)

4. Affected relative or relatives who were diagnosed with heart disease at <55 years of age (recurrence risk ≥2× the general population risk)

E. Cancer (Table 7-1).

1. Cancers are genetic disorders, but most of them are not strictly inherited. The fact that cancers tend to cluster in families demonstrates that there is a genetic component to the disorders as a group.

2. For example, the risk of developing breast cancer doubles for first-degree relatives of women diagnosed with the disorder. The risk also increases if more than one first-degree relative has breast cancer and increases even more if the cancers developed relatively early (before 45 years of age). However, the genetic components of many cancers are poorly understood. The genes known to be involved in hereditary cancers will be discussed in Chapter 16.

3. The role of environmental factors is also recognized in the etiology of many cancers (e.g., the role of tobacco use in lung cancer). The role of infectious agents is also recognized in the etiology of ≈15% of cancers (e.g., the role of human papilloma virus [HPV] in cervical cancer).

t a b l e **7-1** Environmental and Infectious Factors in Cancer	
Factor	**Cancer**
Cigarette smoking (3-methylcholanthrene)	Squamous cell carcinoma of the lung
Asbestos	Pleural mesothelioma
Nitrosamines	Stomach carcinoma
Ultraviolet radiation	Melanoma, basal cell carcinoma, squamous carcinoma of the skin
Nulliparity Oral contraceptive Estrogen replacement therapy	Breast cancer
Low-fiber diet	Colorectal cancer
Nickel, silica, beryllium, chromium	Bronchogenic carcinoma
Benzene	Acute leukemia
Cyclophosphamide, ß-naphthylamine	Transitional cell carcinoma of the urinary bladder
Diethylstilbestrol	Clear cell carcinoma of the vagina
Human papilloma virus (HPV)	Carcinoma of the cervix
Human Herpesvirus 8 (HHV 8)	Kaposi sarcoma
Hepatitis B and C viruses (HBV and HCV)	Primary hepatocellular carcinoma
Human T-cell leukemia virus (HTLV-1)	T-cell leukemia and lymphoma

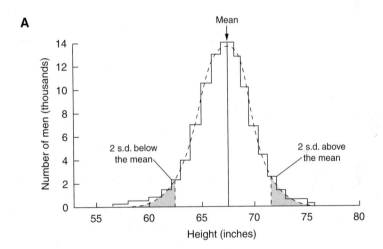

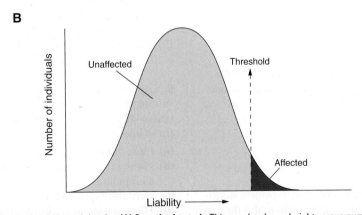

FIGURE 7-1. Classes of multifactorial traits. (A) Quantitative trait. This graphs shows height versus number of men in the population. Note that this quantitative trait follows a normal distribution pattern in the population. **(B) Threshold trait.** This graph shows liability versus number of individuals. Note the liability distribution where affected individuals (*black area*) are seen only above the threshold. Unaffected individuals (*light gray area*).

Review Test

1. Recurrence risk for a multifactorially inherited birth defect is higher in which of the following circumstances?

(A) the proband is of the less commonly affected sex
(B) the proband is less severely affected
(C) only one person in the family is affected
(D) the proband is a first cousin

2. Which of the following best describes the threshold of liability?

(A) maximal risk for a trait with a bell-shaped population distribution
(B) a high discordance rate in dizygotic twin pairs
(C) a low concordance rate in monozygotic twin pairs
(D) minimal level of defect-causing genes and environmental factors for trait occurrence

Use the following pedigree for the next question.

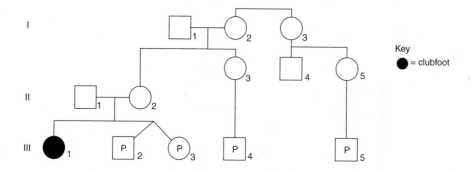

Key
● = clubfoot

3. Isolated clubfoot is more common in males than females and follows multifactorial inheritance. Who has the highest recurrence risk for clubfoot?

(A) II-4
(B) III-2
(C) III-3
(D) III-4
(E) III-5

4. In Ireland, the incidence of neural tube defects is 1 in 200. What is the approximate recurrence risk for first-degree relatives?

(A) 1 in 100
(B) 1 in 37
(C) 1 in 14
(D) 1 in 5

5. A female child is born with pyloric stenosis, which is more common in males than in females. Which of the following best describes the recurrence risk to future siblings?

(A) The recurrence risk for a future brother will increase.
(B) The recurrence risk for a future sister will increase.
(C) The recurrence risk for both a future brother and future sister will increase.
(D) The recurrence risk for both a future brother and future sister will not increase.

6. Concordance for cleft palate is 25% for monozygotic twins versus 10% for dizygotic twins. Which of the following is indicated by these figures?

(A) Heritability for cleft palate is low.
(B) There is a large genetic component to cleft palate.
(C) Heritability for cleft palate is high.
(D) There is a small environmental component to cleft palate.

7. Which one of the following would be expected to increase recurrence risk the most for a multifactorial disorder?

(A) The disorder occurs in a third-degree relative.

(B) There is one affected second cousin.

(C) There is a first cousin with mild disease.

(D) The parents are first cousins.

8. A multifactorial disorder for which the genetic component is thought to be involvement of the HLA (human leukocyte antigen) type II genes and the environmental component is hypothesized to be infection with a virus is which one of the following?

(A) cancer

(B) diabetes Type 1

(C) diabetes Type 2

(D) hypertension

9. There are a number of risk factors in heart disease, but the in the family history the greatest risk would be due to which one of the following?

(A) having a maternal uncle with heart disease

(B) having a paternal uncle with heart disease

(C) having a mother with heart disease

(D) having a father with heart disease

10. Human papilloma virus is an infectious environmental factor in which one of the following multifactorial diseases?

(A) colorectal cancer

(B) hepatocellular cancer

(C) lung cancer

(D) cervical cancer

11. If the heritability (H) of a disorder is entirely due to genetic factors, then H would be equal to which one of the following?

(A) 0

(B) 0.01

(C) 0.10

(D) 1.0

Answers and Explanations

1. **The answer is (A).** When the proband is of the less affected sex, more of the genetic and environmental factors that contribute to the condition are present and subsequent siblings are at a higher risk for the condition because of that. There is also a greater risk to a sibling if the proband is more severely affected, if there are multiple family members affected, or if the affected family member is a first-degree relative.

2. **The answer is (D).** The threshold of liability is reached when the environmental and genetic factors that contribute to a trait reach a level where the trait occurs.

3. **The answer is (B).** Because isolated clubfoot is more common in males than in females, the fact that the daughter III-1 has the condition raises the risk to subsequent siblings. Male siblings will be at a greater risk than female siblings, but the risk is elevated for both. Thus, the male fetus III-2 is at the highest risk for isolated clubfoot. None of the other pregnancies in the family are first-degree relatives so the risk is lower for them.

4. **The answer is (C).** For multifactorial disorders, the risk of recurrence for first-degree relatives is approximately the square root of its incidence.

5. **The answer is (C).** When the affected child in the family is of the less-affected sex, then the recurrence risk for an opposite sex sibling is higher but the risk to a same sex sibling is also increased. This is because if the less-affected sex is affected by a disorder then there are more of the environmental and genetic factors that cause the disorder present, thus lowering the threshold of liability and increasing the recurrence risk.

6. **The answer is (A).** Heritability increases as the concordance between monozygotic twins for a trait increases versus the concordance in dizygotic twins. Because the difference in concordance for cleft palate between monozygotic and dizygotic twins is not very large, heritability is low and there is not a very large genetic component.

7. **The answer is (D).** Consanguinity increases the risks for birth defects in general and confers a high risk of sharing predisposing genes. Having a third-degree relative with a multifactorial disorder does not increase recurrence risk above the population risk and there is only a slight risk with a second-degree relative with mild disease.

8. **The answer is (B).** It is thought that alleles linked to the HLA-DR3 and HLA-DR4 genes alter the immune response such that an immune response to an environmental antigen like a virus gets out of control and destroys the pancreatic beta cells that make insulin. In Type 2 diabetes, there is almost always some insulin production, but individuals with this condition develop insulin resistance. The genetic component of this disease is not well understood.

9. **The answer is (C).** Having a first-degree relative increases recurrence risk for heart disease and having a female first-degree relative (least-affected sex) confers an even higher risk.

10. **The answer is (D).** Human papilloma viruses have been implicated as a factor in the development of cervical carcinoma. A vaccine has been developed to stop infection with the virus and help prevent the development of cervical cancer.

11. **The answer is (D).** If a disorder is entirely due to environmental factors, then H = 0. If the disorder is entirely genetic, then H = 1. A multifactorial disorder would have a value for H somewhere between these two figures because there are both environmental and genetic factors involved in developing the disorder.

8 Population Genetics

I. GENERAL FEATURES

Population genetics is the study of the distributions of genes in populations. In addition, population genetics concerns itself with the factors that maintain or change the frequency of genes (i.e., **gene [allele] frequency**) and frequency of genotypes (i.e., **genotype frequency**) from generation to generation.

A. The **disease frequency** is the frequency that a genetic disorder is observed in a population. It is calculated by data-mining hospital records and is expressed as, for example, 1 in 150,000 people.

B. The **mutation frequency** is the frequency that a mutation occurs in the DNA. The mutation frequency is expressed as the **number of mutations per locus per gamete per generation.**

C. A **gene** is the basic unit of hereditary composed of a finite number of nucleotides arranged in a specific sequence that interacts with the environment to produce a trait.

D. An **allele** is an alternative (or mutated) version of a gene or DNA segment.

E. A **locus** is the physical location of a gene or DNA segment on a chromosome. Since chromosomes are paired in humans, humans have two alleles at each locus.

F. A **polymorphism** is the occurrence of two or more alleles at a specific locus in frequencies greater than can be explained by mutations alone. Polymorphisms are common in noncoding regions of DNA (i.e., introns). A polymorphism does not cause a genetic disorder. A polymorphism can be used as a **genetic marker** for a gene (e.g., the DMD gene on chromosome Xp21.2 for dystrophin related to Duchenne muscular dystrophy) if the polymorphism and the DMD gene are closely linked.

G. Single nucleotide polymorphisms are silent mutations that accumulate in the genome.

H. A **restriction fragment length polymorphism (RFLP)** is a polymorphism that either creates or destroys a restriction enzyme site. RFLPs are abundant throughout the human genome. RFLPs are used in **gene linkage** or **gene mapping** studies. **See Figure 8-1A.**

I. A **variable number tandem repeat (VNTR) polymorphism** is a polymorphism whereby DNA sequences are repeated in tandem in the human genome between restriction enzyme sites a variable number of times. Because VNTRs are extremely polymorphic, two unrelated people cannot exhibit the same genotype. VNTRs are used in **DNA fingerprinting** and **forensic medicine** to establish paternity; zygosity; or identity from a blood, semen, or other DNA sample. In general, VNTRs are better markers than RFLPs. **See Figure 8-1B.**

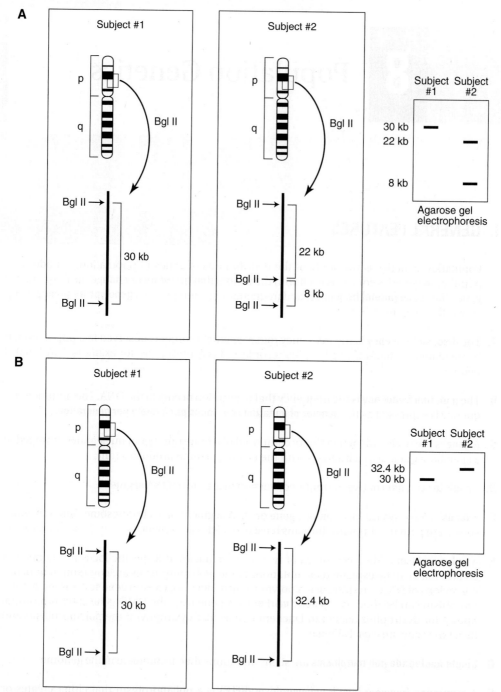

FIGURE 8-1. RFLPs and VNTRs. (A) RFLPs. In subject 1, two Bgl II restriction enzymes sites are present at a certain locus on a chromosome, which produces a 30 kb fragment that is observable with agarose gel electrophoresis. In subject 2, three Bgl II restriction enzyme sites are present at a certain locus on a chromosome, which produces a 22 kb and an 8 kb fragment that is observable with agarose gel electrophoresis. Consequently, subject 2 has a restriction fragment length polymorphism. **(B) VNTRs.** In subject 1, two Bgl II restriction enzymes sites are present at a certain locus on a chromosome, which produces a 30 kb fragment that is observable with agarose gel electrophoresis. In subject 2, two Bgl II restriction enzymes sites are present at a certain locus on a chromosome, which produces a 32.4 kb fragment that is observable with agarose gel electrophoresis. The VNTR has 40 tandem repeats of 60 base pairs ($40 \times 60 = 2,400$ base pairs) inserted to produce the 32.4 kb fragment.

J. A **genotype** is the gene constitution at a specific locus or the specific mutations present in a given disorder.

K. A **phenotype** is the observed physical or clinical findings in a normal person or affected patient, respectively.

II. THE HARDY-WEINBERG LAW

A. Hardy-Weinberg Principles.
1. The Hardy-Weinberg Law explains that the allele frequencies do not change from generation to generation in a large population with random mating.
2. The Hardy-Weinberg Law explains that the genotype frequency is determined by the relative allele frequencies at that locus.
3. The Hardy-Weinberg Law relates the genotype frequency at a locus to the phenotype frequency in a population.

B. Hardy-Weinberg Equation. In a population at equilibrium, for a locus with two alleles (A and a) with allele frequencies of p and q, respectively: the genotype frequencies are determined by the binomial expansion $(p + q = 1)^2$ or $p^2 + 2pq + q^2 = 1$. Therefore, the genotype frequencies are $AA = p^2$, $Aa = 2pq$, and $q^2 = aa$. This is illustrated by the following Punnett square:

	Male Population	
	A (p)	a (q)
Female Population		
A (p)	AA (p^2)	Aa (pq)
a (q)	Aa (pq)	aa (q^2)

C. Hardy-Weinberg Assumptions. The Hardy-Weinberg Law applies if the following assumptions are met:
1. **There is a large population so that there is no influence of genetic drift or the founder effect.**
   a. Genetic drift is a fluctuation in allele frequency **due to chance** operating on a small gene pool contained within a small population. In a small population, random factors such as fertility or survival of mutation carriers can cause the allele frequency to rise for reasons other than the mutation itself.
   b. The founder effect is genetic drift that occurs when one of the founders of a new population carries a rare allele that will have a far higher frequency that it did in the larger population from which the new population is derived. Examples include: **Huntington disorder** in Lake Maracaibo, Venezuela and **variegate porphyria** in Afrikaner populations of South Africa.
2. **There is random mating (panmixia) with no stratification, assortative mating or consanguinity.**
   a. Stratification refers to a population where there are subgroups that have remained, for the most part, genetically distinct. An example would be cystic fibrosis, where the carrier frequency of the disorder is 1/25 in Caucasian, 1/30 in Ashkenazi Jewish, 1/46 in Hispanic, 1/65 in African American, and 1/90 in Asian American populations.
   b. Assortative mating is where the choice of a mate is because of some particular trait. An example of this would be a deaf person preferring another deaf person as a mate.
   c. Consanguinity is where the two mates are related. Progeny from a marriage between first cousins is an example where this would apply.
3. **There is a constant mutation rate where genetically lethal alleles (causing death or sterilization) are replaced by new mutations.** If lethal alleles were not replaced by new mutations, the allele frequency would change with each generation and quickly reach zero.

4. **There is no selection for any of the genotypes at a locus.** Selection acts upon the fitness of a genotype, with **fitness (f)** being a measure of the offspring possessing the genotype that survive to reproduce. A coefficient of selection can be determined, which is $1 - f$.
   a. If the allele determining the genotype is as likely to be present in the next generation as any other allele, then $f = 1$.
   b. If the allele is a genetic lethal, causing death or sterility, then $f = 0$.
   c. If an allele is deleterious so that fewer than normal offspring with the allele are represented in the next generation, then $f < 1$.

5. **There is no migration; no gene flow into or out of the population.** Genes from a migrant population are gradually merged into the gene pool of the population into which they migrated, which can change gene frequencies. Gene flow between populations can slowly change frequencies in both populations making them more similar to each other. For example, the frequency of sickle cell trait is lower in African Americans than in West African populations because of the admixture with other ethnic groups in the United States.

III. HARDY-WEINBERG AND AUTOSOMAL DOMINANT INHERITANCE

In autosomal dominant disorders, homozygosity is exceedingly rare in the population since this is a genetic lethal genotype most of the time. A good example of this is **achondroplastic dwarfism**. Homozygosity for achondroplastic dwarfism is a genetic lethal so that no individuals with this genotype survive to birth. Therefore, individuals in the population with achondroplastic dwarfism are heterozygotes. An exception to this is **Huntington disorder** where homozygosity for Huntington disorder has the same clinical severity as the heterozygosity.

Example 8-1. Achondroplasia Dwarfism
Question: A recent study of achondroplasia dwarfism documented 7 new cases out of 250,000 in the US population. In achondroplasia dwarfism, homozygosity is a genetic lethal. What is the allele frequency of the achondroplasia dwarfism disease gene in the U.S. population?

Solution:
1. 7 new cases out of 250,000 means that achondroplasia dwarfism occurs in the U.S. population at a **disease frequency of 1 in 35,714 or 0.000028** (7/250,000).
2. All the genotypes containing the achondroplasia disease gene include only the heterozygotes (Dd;2pq) since homozygosity is a genetic lethal. Thus, the frequency of heterozygotes (2pq) is equal to the disease frequency (**2pq = 0.000028**).
3. The allele frequency (p) of the disease gene (D) is usually very small in autosomal dominant disorders and the allele frequency (q) of the other allele (d) is ≈ 1. Consequently, 2p(1) or 2p equals the disease frequency. This means that **2p(1) = 0.000028 or 2p = 0.000028 or p = 0.000014**. In conclusion, the allele frequency (p) of the achondroplasia dwarfism disease gene is **0.000014 or 1 in 71,428.**

Example 8-2. Huntington Disorder
Question: Data-mining of the clinical records of a large number of U.S. hospitals indicates that Huntington disorder occurs in the U.S. population at a disease frequency of 1 in 10,000 or 0.0001. In Huntington disorder, homozygosity is not a genetic lethal. What is the allele frequency of the Huntington disease gene in the U.S. population?

Solution:
1. All the genotypes containing the Huntington disease gene include the homozygotes (DD; p^2) and heterozygotes (Dd; 2pq). Thus, the frequency of the homozygotes (DD; p^2) and heterozygotes (Dd; 2pq) is equal to the disease frequency (**p^2 + 2pq = 0.0001**).
2. The frequency of homozygotes (p^2) is very small (≈ 0) and the allele frequency (q) is ≈ 1. Consequently, 2p(1) or 2p equals the disease frequency. This means that **2p(1) = 0.0001 or**

2p = 0.0001 or **p = 0.00005**. In conclusion, the allele frequency (p) of the Huntington disease gene is **0.00005** or **1 in 20,000**.

IV. HARDY-WEINBERG AND AUTOSOMAL RECESSIVE INHERITANCE

A. In autosomal recessive disorders, homozygosity (aa) produces the disorder. A good example of this is **sickle cell anemia**. Homozygosity (aa) for sickle cell anemia produces the disorder in individuals. Therefore, individuals in the population with sickle cell anemia are homozygotes (rr) and the heterozygotes (Rr) are normal but carriers.

Example 8-1. Sickle Cell Anemia
Question: A recent study of sickle cell anemia documented 10 new cases out of 6,250 in the African American population. What is the allele frequency (q) of the sickle cell disease gene in the African American population? What is the allele frequency (p) of the sickle cell normal gene in the African American population? What is the frequency of heterozygote carriers in the African American population?

Solution:
1. 10 new cases out of 6,250 means that sickle cell anemia occurs in the African American population at a **disease frequency of 1 in 625 or 0.0016** (10/6,250).
2. All the genotypes containing the sickle cell disease gene include the homozygotes (rr; q^2) and heterozygotes (Rr; 2pq). The only genotype that produces sickle cell anemia is the homozygote (rr) so that q^2 **= 0.0016**. Taking the square root, gives **q = 0.04**. In conclusion, the allele frequency (q) of the sickle cell disease gene is **0.04** or **1 in 25**.
3. The allele frequency (p) of the sickle cell normal gene is **1-q**. This means that **p = 1.00 − 0.04 = 0.96**. In conclusion, the allele frequency (p) of the sickle cell normal gene is **0.96** or **1 in 1.04**.
4. The frequency of heterozygote carriers is **2pq**. This means that **2pq = 2(0.96)(0.04) = 0.0768**. In conclusion, the frequency of heterozygote carriers is **0.0768** or **1 in 13**.

B. In rare autosomal recessive disorders, the allele frequency (q) is very small and allele frequency (p) is ≈1. In these circumstances, the frequency of heterozygous carriers is approximately equal to 2p.

Example 8-2. Congenital Deafness
Question: A recent study of congenital deafness caused by a connexin 26 mutation documented 1 case out of 4,356 in the U.S. population. What is the allele frequency (q) of the connexin 26 mutation in the U.S. population? What is the allele frequency (p) of the connexin 26 normal gene in the U.S. population? What is the frequency of heterozygote carriers in the U.S. population?

Solution:
1. One case out 4,356 means that congenital deafness caused by the connexin 26 mutation occurs in the U.S. population at a **disease frequency of 1 in 4,356 or 0.00023** (1/4356).
2. All the genotypes containing the connexin 26 mutation include the homozygotes (rr;q^2) and heterozygotes (Rr; 2pq). The only genotype that produces congenital deafness is the homozygote (rr) so that q^2 **= 0.00023**. Taking the square root gives **q = 0.0151**. In conclusion, the allele frequency (q) of the connexin 26 mutation is **0.0151** or **1 in 66**.
3. The allele frequency (p) of the connexin 26 normal gene is **1 − q**. This means that **p = 1 − 0.0151 = 0.9848 (or ≈1)**. In conclusion, the allele frequency (p) of the connexin 26 normal gene is **0.9849** or **1 in 1.02**.

4. The frequency of heterozygote carriers is **2pq**. This means that **2pq = 2(1) (0.0151) = 0.0302**. In conclusion, the frequency of heterozygote carries for the connexin 26 mutation is **0.0302** or **1 in 33**.

C. Because only a small proportion of recessive alleles are present in homozygotes selection does not have much affect on allele frequencies. Even if $f = 0$, it would take many generations to reduce a mutant allele frequency appreciably. In some cases, the heterozygotes have a selective advantage. For example, heterozygous carriers of sickle cell trait have resistance to malaria. Sickle cell trait has its highest carrier frequency in West Africa where malaria is prevalent.

V. HARDY-WEINBERG AND X-LINKED DOMINANT INHERITANCE

In X-linked dominant disorders, the disorder is observed in twice the number of females than males. Males usually die (a genetic lethal). In X-linked dominant disorders, females who receive the mutant gene on the X chromosome have the disorder. The genotype of a normal female is XX and the genotype of an affected female is X^DX. A good example of this is classic Rett syndrome.

Example: Classic Rett Syndrome
Question: A recent study of classic Rett syndrome documented 10 new cases out of 180,000 over the last 10 years in the female U.S. population. What is the allele frequency (q) of the classic Rett syndrome disease gene in the U.S. population? What is the allele frequency (p) of the classic Rett syndrome normal gene in the U.S. population?

Solution:
1. 10 new cases out of 180,000 means that classic Rett syndrome occurs in the U.S. population at a **disease frequency of 1 in 18,000 or 0.00006** (1/18,000).
2. In X-linked dominant disorders, the allele frequency (q) of the classic Rett syndrome disease gene equals the disease frequency. This means that **q = 0.00001**. In conclusion, the allele frequency (q) of the classic Rett syndrome disease gene is **0.00001 or 1 in 100,000**.
3. The allele frequency (p) of the classic Rett syndrome normal gene is **1 − q**. This means that **p = 1.00 − 0.00001 = 0.99999 (or ≈1)**. In conclusion, the allele frequency (p) of the classic Rett syndrome normal gene is **0.99999 or 1 in 1.00001**.

VI. HARDY-WEINBERG AND X-LINKED RECESSIVE INHERITANCE

In X-linked recessive disorders, the disorder is observed only in males (affected homozygous females are rare). In X-lined recessive disorders, males who receive the mutant gene on the X chromosome have the disorder. The genotype of a normal male is XY and the genotype of an affected male is X^rY. A good example of this is **Hunter syndrome (mucopolysaccharidosis II)**.

Example: Hunter Syndrome
Question: A recent study of Hunter syndrome documented 10 new cases out of 1,000,000 over the last 5 years in the U.S. population. What is the allele frequency (q) of the Hunter syndrome disease gene in the U.S. population? What is the allele frequency (p) of the Hunter syndrome normal gene in the U.S. population? What is the frequency of female heterozygote carriers?

Solution:
1. 10 new cases out of 1,000,000 means that Hunter syndrome occurs in the U.S. population at a **disease frequency of 1 in 100,000 or 0.00001** (1/100,000).

2. In X-linked recessive disorders, the allele frequency (q) of the Hunter syndrome disease gene equals the disease frequency. This means that **q = 0.00001.** In conclusion, the allele frequency (q) of the Hunter syndrome disease gene is **0.00001 or 1 in 100,000.**
3. The allele frequency (p) of the Hunter syndrome normal gene is **1 − q.** This means that **p = 1.00 − 0.00001 = 0.99999 (or ≈1).** In conclusion, the allele frequency (p) of the Hunter syndrome normal gene is **0.99999 or 1 in 1.00001.**
4. The frequency of female heterozygote carriers is **2pq,** where p = ≈1. This means that **2pq = 2 (1) (0.00001) = 0.00002.** In conclusion, the frequency of female heterozygote carriers is **0.00002 or 1 in 50,000 females.**

VII. MUTATION-SELECTION EQUILIBRIUM

According to the mutation-selection equilibrium theory, gene frequencies in a population are maintained in equilibrium by mutations that replace disease genes that are selectively lost in the population by death (e.g., a genetic lethal disease) or reproduction failure (e.g., sterility).

A. **Autosomal Dominant Disorders.** If every carrier of a disease gene dies or cannot reproduce, then every new case of the autosomal dominant disorder arises from a new mutation.
   1. The **disease frequency** is the frequency that a genetic disorder is observed in a population. The disease frequency is expressed as, for example, 1 in 150,000 people. Therefore, if the number of affected people in the population studied is **D** and number of people in the population studied is **N,** then the **disease frequency = D/N.**
   2. The **mutation frequency** is the frequency that a mutation occurs in the DNA. The mutation frequency is expressed as the number of new mutations per locus or the number of new mutations per gamete. Therefore, if the number of affected people in the population studied is **D** and number of people in the population studied is **N,** then the **mutation frequency = D/2N.** The denominator is 2N because a mutation of either allele at the autosomal locus could result in the disorder.

   ### Example: Achondroplasia Dwarfism
   Question: A recent study of achondroplasia dwarfism documented 7 new cases out of 250,000 in the U.S. population. In achondroplasia dwarfism, homozygosity is generally a genetic lethal. What is the disease frequency? What is the mutation frequency?

   Solution:
   1. The disease frequency = D/N = 7/250,000 = 0.000028 = 2.8×10^{-5} = 0.0028%. Dividing 100/0.0028 = 35,714. Therefore, the disease frequency is 0.000028, 2.8×10^{-5}, 0.0028%, or 1 in 35,714 people.
   2. The mutation frequency = D/2N = 7/2(250,000) = 0.000014 = 1.4×10^{-5} = 0.0014%. Dividing 100/0.000014 = 71,428. Therefore, the mutation frequency is 0.000014, 1.4×10^{-5}, 0.0014%, or 1 new mutation per 71,428 loci or 1 new mutation per 71,428 gametes.

B. **X-linked Recessive Disorders.** If every carrier of a disease gene dies or cannot reproduce, then every new case of the X-linked recessive disorder arises from a new mutation.
   1. The **disease frequency** is the frequency that a genetic disorder is observed in the population. The disease frequency is expressed as, for example, 1 in 150,000 people. Therefore, if the number of affected people in the population studied is **D** and number of people in the population studied is **N,** then the **disease frequency = D/N.**
   2. The **mutation frequency** is the frequency that a mutation occurs in the DNA. The mutation frequency is expressed as the number of new mutations per locus or the number of new mutations per gamete. Therefore, if the number of affected people in the population studied is **D** and number of people in the population studied is **N,** then the **mutation frequency = D/3N.** The denominator is 3N because in a population of equal numbers of females and males, only 1/3 of the X chromosomes occur in the male. An X-linked recessive disorder is observed only in males. Heterozygous females are generally clinically normal.

Example: Fragile X Syndrome

Question: In a recent study of fragile X syndrome, 15 new cases in boys were documented among 250,000 births. What is the disease frequency? What is the mutation frequency?

Solution:
1. The disease frequency = D/N = 15/250,000 = 0.00006 = 6.0×10^{-5} = 0.006%. Dividing 100/0.006 = 16,666. Therefore, the disease frequency is 0.00006, 6.0×10^{-5}, 0.006%, or 1 in 16,666 people.
2. The mutation frequency = D/3N = 15/3(250,000) = 0.00002 = 2.0×10^{-5} = 0.002%. Dividing 100/0.002 = 50,000. Therefore, the mutation frequency is 0.00002, 2.0×10^{-5}, 0.002%, or 1 new mutation per 50,000 loci or 1 new mutation per 50,000 gametes.

VIII. LINKAGE

Linkage is the closeness of two or more loci on a chromosome that sort together during meiosis. Linkage refers to loci, not alleles. Linkage occurs because crossover rarely occurs between loci that are close together. **Crossover** is the exchange of DNA between homologous chromosomes that occurs in meiosis I during **chiasma (i.e., the crossover point)** formation. Linkage is measured only in family pedigrees because meiosis that is required to demonstrated linkage occurs only during gametogenesis. Consequently, only family members can be used to determine linkage. The units of measurement of linkage include the following:

A. **Recombination Fraction (θ) or Recombination %.** If the crossover point does *not* occur between the two loci under consideration, then a **parental chromosome** is formed. If the crossover point does occur between the two loci under consideration, then a **recombinant chromosome** is formed. This is measured by the recombination fraction (θ) and the recombination % and is best illustrated by the example below.

   Example: A mother and father have a family of 8 children. A chromosomal analysis revealed that 7 children had parental chromosomes, which means the crossover point did not occur between the two loci under consideration during meiosis. However, 1 child had a recombinant chromosome, which means the crossover point occurred between the two loci under consideration during meiosis. What is the recombinant fraction (θ)? What is the recombination %?

$$\text{Recombination Fraction } (\theta) = \frac{\text{\# of recombinant chromosomes}}{\text{total number of children}} = 1/8 = 0.13$$

$$\text{Recombination \%} = \frac{\text{\# of recombinant chromosomes}}{\text{total number of children}} \times 100 = 1/8 = 0.13 \times 100 = 13\%$$

B. **Centimorgan (cM).** cM defines the distance between two loci. 1cM is defined as the distance between two loci that produces a recombination fraction (θ) = 0.01 or a recombination % = 1%.
   1. **High linkage.** If two loci are **0.1cM apart**, then the **recombinant fraction (θ) = 0.001**, and the **recombinant % = 0.1%.** This means there is a **0.1% chance** (i.e., 1 in a 1,000 chance) of recombination occurring between these loci during crossover. Therefore, if a 1,000 gametes are produced during gametogenesis, 1 gamete will contain a recombinant chromosome.
   2. **Intermediate linkage.** If two loci are **1cM apart**, then the **recombinant fraction (θ) = 0.01**, and the **recombinant % = 1.0%.** This means there is a **1% chance** (i.e., 1 in a 100 chance) of recombination occurring between these loci during crossover. Therefore, if a 100 gametes are produced during gametogenesis, 1 gamete will contain a recombinant chromosome.
   3. **Low linkage.** If two loci are **50cM apart**, then the **recombinant fraction (θ) = 0.50**, and the **recombinant % = 50%.** This means there is a **50% chance** (i.e., 50 in 100 chance) of recom-

bination occurring between these loci during crossover. Therefore, if a 100 gametes are produced during gametogenesis, 50 gametes will contain a recombinant chromosome.

C. Logarithm of the Odds (LOD) Score (or Z).

$$\text{LOD} = \log_{10} \frac{\text{probability that 2 loci are linked}}{\text{probability that 2 loci are not linked}}$$

A LOD score of $\geq +3$ indicates definite evidence FOR linkage between two loci.
A LOD score of ≤ -2 indicates definite evidence AGAINST linkage between two loci

---

table **8-1** Summary Table of Population Genetics

	Hardy-Weinberg Calculations **Important Equations: $p^2 + 2pq + q^2 = 1$** **$p + q = 1$**
Autosomal dominant inheritance	Allele frequency (p) of disease gene (D) = $\dfrac{\text{disease frequency}}{2}$ Allele frequency (q) of normal gene (d) = $1 - p$
Autosomal recessive inheritance	Allele frequency (q) of disease gene (r) = $\sqrt{\text{disease frequency}}$ Allele frequency (p) of normal gene (R) = $1 - q$ Frequency of heterozygote carriers = $2pq$
X-linked dominant inheritance	Allele frequency (q) of disease gene (X^D) = disease frequency Allele frequency (p) of normal gene (X^d) = $1 - q$
X-linked recessive inheritance	Allele frequency (q) of disease gene (X^r) = disease frequency Allele frequency (p) of normal gene (X^R) = $1 - q$ Frequency of heterozygote carriers = $2pq$
	Mutation-Selection Equilibrium
Autosomal dominant disorders	Disease frequency = D/N Mutation frequency = D/2N
X-linked recessive Disorders	Disease frequency = D/N Mutation frequency = D/3N
	Linkage Recombination fraction (θ) = $\dfrac{\text{\# of recombinant chromosomes}}{\text{total number of children}}$ Recombination % = $\dfrac{\text{\# of recombinant chromosomes}}{\text{total number of children}} \times 100$ $1\text{cM} = 0.01\ \theta = 1\%$ recombination % $\text{LOD} = \log_{10} \dfrac{\text{Probability that 2 loci are linked}}{\text{Probability that 2 loci are not linked}}$

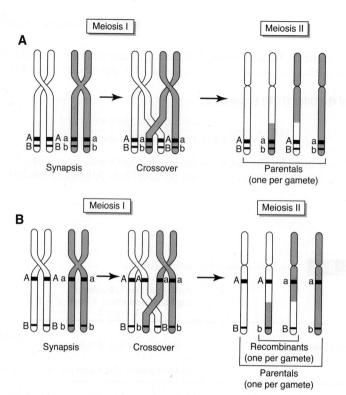

C

Linkage		cM	Recombination Fraction (θ) Recombination %	LOD Score
High		0.1 cM	0.001 θ 0.1% 1 gamete per 1000 gametes will contain a recombinant chromosome	≥ + 3
Intermediate		1 cM	0.01 θ 1.0% 1 gamete per 100 gametes will contain a recombinant chromosome	
Low		50 cM	0.50 θ 50% 50 gametes per 100 gametes will contain a recombinant chromosome	≤ −2

FIGURE 8-2. Linkage. (A) Absolute linkage (0cM, 0.00 θ, 0%, LOD score ≥+3). During meiosis I in gamete formation , two homologous chromosomes undergo synapsis (pairing of homologous chromosomes) and crossover. In this case, the loci are so close together that a crossover point will never occur between the alleles (A, B, a, b). Therefore, only parental chromosomes **(parentals)** will form. During meiosis II, the gametes that are formed will have the same allele pattern under consideration as the parents. **(B) Low linkage (50cM, 0.50 θ, 50%, LOD score ≤−2).** During meiosis I in gamete formation, two homologous chromosomes undergo synapsis (pairing of homologous chromosomes) and crossover. In this case, the loci are very far apart so that a crossover point will occur between the alleles (A, B, a ,b). Therefore, both parental chromosomes **(parentals)** and recombinant chromosomes **(recombinants)** will form. During meiosis II, the 50% of the gametes that are formed will have the same allele pattern under consideration as the parents and 50% of the gametes that are formed will have different allele pattern under consideration than the parents. **(C) Measurements of linkage.** Examples of high, intermediate, and low linkage are shown.

Review Test

1. In an autosomal recessive disease, the people represented by the "2pq" figure in the equation "$p^2 + 2pq + q^2$" are:

(A) carriers of the disease
(B) affected by the disease
(C) those who have a new mutation
(D) those without the mutant gene

2. Duchenne muscular dystrophy is a lethal X-linked recessive disease, which affects 1 in 3,500 boys. What is the carrier frequency of this gene mutation in females?

(A) 1/3,500
(B) 1/1,750
(C) 1/59
(D) 3/50
(E) 1/25

3. A study was undertaken of pregnant women in North Carolina to determine the ratios of women with sickle cell disease and sickle cell trait. The study found that 60% of the African American women in North Carolina are homozygous for the normal dominant hemoglobin allele. What percentage of the women would be carriers of sickle cell trait?

(A) 5%
(B) 25%
(C) 35%
(D) 40%

4. Huntington disease is an autosomal dominant disease in which those who are homozygous for the disease gene have a clinical course that is no different from that of heterozygotes. In a population where the frequency of the Huntington gene is 0.08, what is the frequency of those who are homozygous for the gene?

(A) 0.064
(B) 0.016
(C) 0.08
(D) 0.0064

5. In achondroplastic dwarfism, an autosomal dominant disease, the gene is lethal in homozygotes. If the frequency of the normal

allele is 0.99, what is the heterozygote frequency?

(A) 0.98
(B) 0.01
(C) 0.02
(D) 0.99

6. In South Africa, variegate porphyria is found in white South Africans at a higher frequency than would be expected if the population was in Hardy-Weinberg equilibrium. This population originated from a small group of Dutch settlers. The most likely explanation for the high frequency of variegate porphyria in this population is:

(A) selection for heterozygotes
(B) the founder effect
(C) immigration into the population
(D) selection against heterozygotes

7. It is believed that the cystic fibrosis (CF) gene conferred protection to carriers during the cholera epidemics of the Middle Ages. The CF gene frequency would be expected to do what in those populations?

(A) increase
(B) decrease
(C) stay the same
(D) become sex-linked recessive

8. Early in the 20th century, eugenics proponents thought that if those with genetic diseases were prevented from having children, then the diseases would disappear. They were erroneous in their reasoning, however. Why wouldn't this strategy work for autosomal recessive diseases?

(A) because only females are carriers
(B) because only males are carriers
(C) because the genes would be "protected" in carriers
(D) because 1/4 of all children would be normal

9. Hemochromatosis is an autosomal recessive disease that is relatively common in the population (1 in 500). The disease can be treated successfully by periodic removal of blood (serial phlebotomy) but failure to recognize it can lead to a number of serious conditions. Population screening for carriers is being considered. What is the expected carrier frequency in the population?

(A) 0.10%
(B) 0.20%
(C) 2.25%
(D) 8.5%
(E) 15%

10. The recombinant % of two loci is 66%. Which one of the following is the explanation for this figure?

(A) The two loci are far apart and there is low linkage between them.
(B) The two loci are close together and there is high linkage between them.
(C) The recombinant fraction between the loci is low.
(D) The LOD score is high.

11. A couple has had four children, two boys and two girls. The father is a carrier of a pericentric inversion of chromosome 8. Two of the couple's children were born with recombinant chromosomes and died shortly after birth. What is the recombinant fraction for the inversion?

(A) 0.10
(B) 0.25
(C) 0.50
(D) 0.75

12. Which one of the following violates the assumptions upon which the Hardy-Weinberg Law is based?

(A) There is no selection against any allele in the population.
(B) There is a constant mutation rate where lethal genes are replaced by new mutations.
(C) There is a large population with assortative mating.
(D) There is no migration into the population.

13. An autosomal recessive gene is lethal in homozygotes. In a population at Hardy-Weinberg equilibrium, how many generations will it take to eliminate the gene from the population

(A) 1
(B) 25
(C) 100
(D) It will not be eliminated from the population.

14. In West Africa, the sickle cell gene (S) frequency is 0.15, while in the United States the frequency of (S) is about 0.04. The most likely explanation for this is which one of the following?

(A) selection for homozygotes
(B) selection against homozygotes
(C) selection for heterozygotes
(D) selection against heterozygotes

15. If a dominant gene is lethal (fitness = 0), why do you continue to see it in a population?

(A) some individuals are carriers
(B) it is "protected" in homozygotes
(C) it is "protected" in heterozygotes
(D) it continues to arise as a new mutation

16. The frequency of the autosomal recessive disease phenylketonuria is 1 in 10,000. What is the carrier frequency for this disease?

(A) 1/50
(B) 1/100
(C) 1/500
(D) 1/1,000

17. Which one of the following best describes the gene frequency for an X-linked recessive disease?

(A) it is equal to 1
(B) it is equal to 2pq − 1
(C) it is equal to its frequency in female carriers
(D) it is equal to the disease frequency

Answers and Explanations

1. **The answer is (A).** In an autosomal recessive disease, those individuals who are represented by "2pq" are heterozygotes who carry a copy of a mutated disease-causing gene.

2. **The answer is (B).** The gene frequency in X-linked recessive diseases is the same as the disease frequency. In this case, the gene frequency is 1/3,500. The carrier frequency is 2pq, since p is almost equal to 1 the carrier frequency is $2 \times 1 \times 1/3,500 = 2/3,500$ or 1/1,750.

3. **The answer is (C).** Because 60% of the women are homozygous for the normal dominant hemoglobin allele, that means that $p^2 = 0.60$ and the square root, or "p" is equal to 0.77. That means that 1 – p, or "q", the frequency of sickle cell trait, is equal to 0.23. The percentage of heterozygotes, or carriers of the sickle cell trait, is thus 2pq, or $2 \times 0.77 \times 0.23$, which is equal to 0.35 or 35%.

4. **The answer is (D).** Because the frequency p of the disease gene is 0.08, then the homozygote frequency would be p^2 or 0.0064.

5. **The answer is (C).** Because the frequency of the normal allele "q" is 0.99, the frequency of the mutated allele "p" is 1 – q or 1 – 0.99, which is 0.01. The frequency of "p" is almost equal to 1, so the heterozygote frequency 2pq is thus $2 \times 1 \times 0.01$ or 0.02.

6. **The answer is (B).** In a small population, a greater proportion of "founding" individuals may carry a gene than in the larger population. The small group of Dutch settlers in South Africa reproduced only with members of the group. If the autosomal dominant mutant gene that causes variegate porphyria was overrepresented in that small population, then a larger proportion of individuals with the gene would be born into the group, and the frequency of the gene would increase.

7. **The answer is (A).** If heterozygosity for the CF gene conferred protection against cholera, then heterozygotes would probably survive to reproduce. Those without the CF gene would most likely die before producing any more children or die before they reached reproductive age. Those who were homozygous for the gene had CF and died without reproducing. Carriers would survive to reproduce, the CF gene would be passed on, and the gene frequency would increase due to the positive selective pressure on the gene.

8. **The answer is (C).** In autosomal recessive diseases, carriers of the disease gene are generally asymptomatic, so it is not obvious who does and who does not carry the gene. Since those who are carrying the gene cannot be identified, the gene is "protected" from removal from the population. Even if that could be accomplished, new mutations would occur that would keep the gene in the population.

9. **The answer is (D).** The frequency of those with the disease, 1 in 500, is equal to 0.02, which is q^2. The frequency of the disease gene "q" is the square root of 0.02, or 0.0447. The carrier frequency is 2pq, and "p" is close to 1, but with the disease being so common, a more accurate carrier frequency can be obtained by using the true value of "p", which is 1 – q = 1 – 0.0447 = 0.9553. So 2pq = 2(0.9553)(0.0447) = 0.0849 which is about 8.5%.

10. **The answer is (A).** The recombinant percent of the two loci is very large, meaning that there is a lot of recombination between them. The further apart two loci are, the higher the recombination percent, so the two loci are far apart and not very tightly linked.

11. **The answer is (C).** Of the couple's four children, two of the children had a recombinant chromosome resulting from crossing over in the inversion loop during paternal meiosis. The recombinant fraction is the number of recombinant chromosomes divided by the number of children. There were 2 recombinant chromosomes in the 4 children, which is 2/4 or 1/2, which is 0.50.

12. **The answer is (C).** Assortative mating, where "like mates with like" violates the assumption that there is random mating in the population. Assortative mating would cause an increase in the frequency of the gene (or genes) responsible for whatever characteristic was associated with the assortative mating, like tall with tall, short with short, etc.

13. **The answer is (D).** The gene will not be eliminated from the population because it is "protected" in heterozygotes who are not affected and can pass the gene on to succeeding generations.

14. **The answer is (C).** Heterozygotes for the sickle cell gene (S) have resistance to malaria. West Africa is a malaria prone area and having resistance to malaria would enhance an individual's chances of living to reproduce. Thus, the sickle cell gene would be "selected for" in the population and its frequency would gradually increase because those with the gene would have more reproductive success.

15. **The answer is (D).** A lethal dominant gene appears in the population as the result of new mutations. There are no "carriers" who are unaffected; it is lethal to all who inherit it.

16. **The answer is (A).** The frequency of phenylketonuria, $1/10,000$ is q^2. So "q" is the square root of $1/10,000$ or $1/100$. Because "p" is close to 1, the carrier frequency, $2pq$ is $2(1)(1/100) = 2/100 = 1/50$, so 1 in 50 is the carrier frequency.

17. **The answer is (D).** The gene frequency for X-linked recessive diseases is the same as the frequency of affected males, which is the disease frequency because females are generally not affected.

chapter 9

Mitosis, Meiosis, and Gametogenesis

I. MITOSIS (Figure 9-1)

- Mitosis is the process by which a cell with the diploid number of chromosomes, which in humans is 46, passes on the diploid number of chromosomes to daughter cells.
- The term **diploid** is classically used to refer to a cell containing 46 chromosomes.
- The term **"haploid"** is classically used to refer to a cell containing 23 chromosomes.
- Mitosis ensures that the diploid number of 46 chromosomes is maintained in cells.
- Mitosis occurs at the end of a cell cycle. Phases of the cell cycle are:

A. **G_0 (Gap) Phase.** The G_0 phase is the resting phase of the cell. The amount of time a cell spends in G_0 is variable and depends on how actively a cell is dividing.

B. **G_1 Phase.** The G_1 phase is the gap of time between mitosis (M phase) and DNA synthesis (S phase). The G_1 phase is the phase where **RNA, protein, and organelle synthesis** occurs. The G_1 phase lasts about **5 hours** in a typical mammalian cell with a 16-hour cell cycle.

C. **G_1 Checkpoint. Cdk2-cyclin D and Cdk2-cyclin E** mediate the **$G_1 \rightarrow$ S phase** transition at the **G_1 checkpoint.**

D. **S (Synthesis) Phase.** The S phase is the phase where **DNA synthesis** occurs. The S phase lasts about **7 hours** in a typical mammalian cell with a 16-hour cell cycle.

E. **G_2 Phase.** The G_2 phase is the gap of time between DNA synthesis (S phase) and mitosis (M phase). The G_2 phase is the phase where high levels of **ATP synthesis** occur. The G_2 phase lasts about **3 hours** in a typical mammalian cell with a 16-hour cell cycle.

F. **G_2 Checkpoint. Cdk1-cyclin A and Cdk1-cyclin B** mediate the **$G_2 \rightarrow$ M phase** transition at the **G_2 checkpoint.**

G. **M (Mitosis) Phase.** The M phase is the phase where **cell division** occurs. The M phase is divided into six stages called **prophase, prometaphase, metaphase, anaphase, telophase,** and **cytokinesis.** The M phase lasts about 1 hour in a typical mammalian cell with a 16-hour cell cycle.
   1. **Prophase.** The chromatin condenses to form well-defined chromosomes. Each chromosome has been duplicated during the S phase and has a specific DNA sequence called the **centromere** that is required for proper segregation. The **centrosome complex,** which is the **microtubule organizing center (MTOC),** splits into two and each half begins to move to opposite poles of the cell. The **mitotic spindle** (microtubules) forms between the centrosomes.

2. **Prometaphase.** The nuclear envelope is disrupted, which allows the microtubules access to the chromosomes. The nucleolus disappears. The **kinetochores** (protein complexes) assemble at each centromere on the chromosomes. Certain microtubules of the mitotic spindle bind to the kinetochores and are called **kinetochore microtubules.** Other microtubules of the mitotic spindle are now called **polar microtubules** and **astral microtubules.**

3. **Metaphase.** The chromosomes align at the metaphase plate. The cells can be arrested in this stage by microtubule inhibitors (e.g., colchicine). The cells arrested in this stage can be used for **karyotype analysis.**

4. **Anaphase.** The centromeres split, the kinetochores separate, and the chromosomes move to opposite poles. The kinetochore microtubules shorten. The polar microtubules lengthen.

5. **Telophase.** The chromosomes begin to decondense to form chromatin. The nuclear envelope re-forms. The nucleolus reappears. The kinetochore microtubules disappear. The polar microtubules continue to lengthen.

6. **Cytokinesis.** The cytoplasm divides by a process called **cleavage.** A **cleavage furrow** forms around the middle of the cell. A **contractile ring** consisting of actin and myosin filaments is found at the cleavage furrow.

II. CHECKPOINTS

A. The checkpoints in the cell cycle are specialized signaling mechanisms that regulate and coordinate the cell response to **DNA damage** and **replication fork blockage.**

- When the extent of DNA damage or replication fork blockage is beyond the steady-state threshold of DNA repair pathways, a checkpoint signal is produced and a checkpoint is activated.
- The activation of a checkpoint slows down the cell cycle so that DNA repair may occur and/or blocked replication forks can be recovered.

B. The two main protein families that control the cell cycle are **cyclins** and the **cyclin-dependent protein kinases (Cdks).**

- A cyclin is a protein that regulates the activity of Cdks and is named because cyclins undergo a cycle of synthesis and degradation during the cell cycle.
- The cyclins and Cdks form complexes called **Cdk-cyclin complexes.**
- The ability of Cdks to phosphorylate target proteins is dependent on the particular cyclin that complexes with it.

III. MEIOSIS (Figure 9-2)

- Meiosis is the process of **germ cell division** (contrasted with mitosis which is **somatic cell division**) that occurs only in the production of the germ cells (i.e., sperm in the testes and oocyte in the ovary).
- In general, meiosis consists of two cell divisions (meiosis I and meiosis II) but only one round of DNA replication that results in the formation of four gametes, each containing half the number of chromosomes (23 chromosomes) and half the amount of DNA (1N) found in normal somatic cells (46 chromosomes, 2N).
- The various aspects of meiosis compared to mitosis are given in Table 9-1.

A. **Meiosis I.** Events that occur during meiosis I include:
1. **DNA replication**
2. **Synapsis.** Synapsis refers to the pairing of each duplicated chromosome with its homologue, which occurs only in meiosis I (not meiosis II or mitosis).

> **a.** In female meiosis, each chromosome has a homologous partner so the two X chromosomes synapse and crossover just like the other pairs of homologous chromosomes.
> **b.** In male meiosis, there is a problem because the X and Y chromosomes are very different. However, the X and Y chromosomes do pair and crossover. The pairing of the X and Y chromosomes is in an **end-to-end fashion** (rather than along the whole length as for all the other chromosomes), which is made possible by a 2.6 Mb region of sequence homology between the X and Y chromosomes at the tips of their p arms where crossover occurs. This region of homology is called the **pseudoautosomal region.**
> **c.** Although the X and Y chromosomes are not homologs, they are functionally homologous in meiosis so there are 23 homologous pairs of the 46 duplicated chromosomes in the cell at this point.

3. **Crossover.** Crossover refers to the **equal exchange** of large segments of DNA between the maternal chromatid and paternal chromatid (i.e., nonsister chromatids) at the **chiasma,** which occurs during prophase (pachytene stage) of meiosis I. Chiasma is the location where crossover occurs forming an X-shaped chromosome and named for the Greek letter *chi*, which also is X-shaped.
   > **a.** Crossover introduces **one level of genetic variability** among the gametes.
   > **b.** During crossover, two other events (i.e., **unequal crossover** and **unequal sister chromatid exchange**) may occur, which introduces **variable number tandem repeat (VNTR) polymorphisms, duplications, or deletions** into the human nuclear genome.

4. **Alignment.** Alignment refers to the process whereby homologous duplicated chromosomes align at the metaphase plate. At this stage, there are still 23 pairs of the 46 chromosomes in the cell.

5. **Disjunction.** Disjunction refers to the separation of the 46 maternal and paternal duplicated chromosomes in the 23 homologous pairs from each other into separate secondary gametocytes (Note: the **centromeres do not split**).
   > **a.** The choice of which maternal or paternal homologous duplicated chromosomes enters the secondary gametocyte is a **random distribution.**
   > **b.** There are 2^{23} **(or 8.4 million)** possible ways the maternal and paternal homologous duplicated chromosomes can be combined. This random distribution of maternal and paternal homologous duplicated chromosomes introduces **another level of genetic variability** among the gametes.

6. **Cell division.** Meiosis I is often called the **reduction division,** because the number of chromosomes is reduced by half, to the haploid (23 duplicated chromosomes, 2N DNA content) number in the two secondary gametocytes that are formed.

B. Meiosis II. Events that occur during meiosis II include:
   1. **Synapsis:** absent.
   2. **Crossover:** absent.
   3. **Alignment:** 23 duplicated chromosomes align at the metaphase plate.
   4. **Disjunction:** 23 duplicated chromosomes separate to form 23 single chromosomes when the **centromeres split.**
   5. **Cell division:** gametes (23 single chromosomes, 1N) are formed.

IV. OOGENESIS: FEMALE GAMETOGENESIS

A. Primordial germ cells (46,2N) from the wall of the yolk sac arrive in the ovary at **week 4** and differentiate into **oogonia (46,2N)** which populate the ovary through *mitotic* division.

B. Oogonia enter meiosis I and undergo DNA replication to form **primary oocytes (46,4N).** All primary oocytes are formed by the **month 5 of fetal life.** No oogonia are present at birth. Primary oocytes remain **dormant in prophase (diplotene) of meiosis I** from month 5 of fetal life until puberty at ≈12 years of age (or ovulation at ≈50 years of age, given that some primary oocytes will remain dormant until menopause).

C. After puberty, 5 to 15 primary oocytes will begin maturation with each ovarian cycle, with usually only one reaching full maturity in each cycle.

D. During the ovarian cycle, a primary oocyte completes meiosis I to form two daughter cells: the **secondary oocyte (23 chromosomes, 2N amount of DNA)** and the **first polar body,** which degenerates.

E. The secondary oocyte promptly begins meiosis II but is **arrested in metaphase of meiosis II** about 3 hours before ovulation. The secondary oocyte remains arrested in metaphase of meiosis II until fertilization occurs. **F.** At fertilization, the secondary oocyte will complete meiosis II to form a **one mature oocyte (23,1N)** and a **second polar body.**

V. SPERMATOGENESIS: MALE GAMETOGENESIS IS CLASSICALLY DIVIDED INTO 3 PHASES

A. Spermatocytogenesis. Primordial germ cells (46,2N) form the wall of the yolk sac, arrive in the testes at **week 4,** and remain **dormant until puberty.** At puberty, primordial germ cells differentiate into **Type A spermatogonia (46,2N).** Type A spermatogonia undergo mitosis to provide a continuous supply of stem cells throughout the reproductive life of the male. Some Type A spermatogonia differentiate into **Type B spermatogonia (46,2N).**

B. Meiosis. Type B spermatogonia enter meiosis I and undergo DNA replication to form **primary spermatocytes** (46,4N). Primary spermatocytes complete meiosis I to form **secondary spermatocytes (23,2N).** Secondary spermatocytes complete meiosis II to form **four spermatids (23,1N).**

C. Spermiogenesis. Spermatids undergo a **postmeiotic series of morphological changes** to form **sperm (23,1N).** These changes include formation of the acrosome; condensation of the nucleus; and formation of head, neck, and tail.
1. The total time of sperm formation (from spermatogonia to spermatozoa) is about 64 days.
2. Newly ejaculated sperm are incapable of fertilization until they undergo **capacitation,** which occurs in the female reproductive tract and involves the unmasking of sperm glycosyltransferases and removal of proteins coating the surface of the sperm. Capacitation occurs before the acrosome reaction.

VI. COMPARISON TABLE OF MEIOSIS AND MITOSIS (Table 9-1)

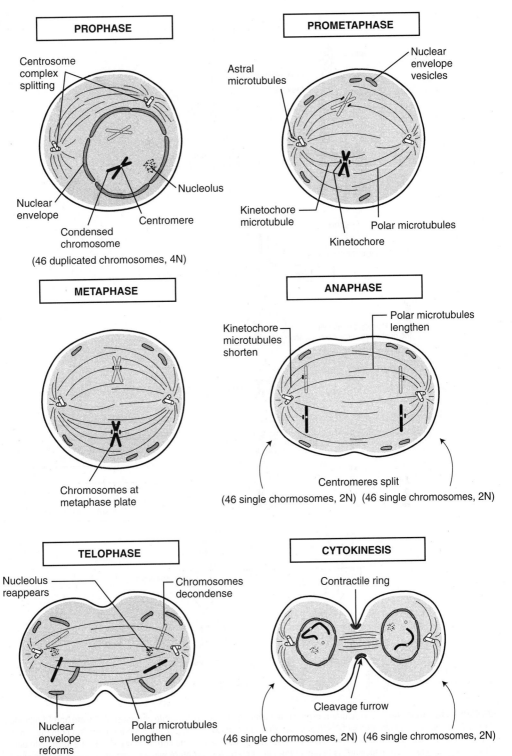

PROPHASE

Centrosome complex splitting

Nuclear envelope

Condensed chromosome

Nucleolus

Centromere

(46 duplicated chromosomes, 4N)

PROMETAPHASE

Astral microtubules

Nuclear envelope vesicles

Kinetochore microtubule

Polar microtubules

Kinetochore

METAPHASE

Chromosomes at metaphase plate

ANAPHASE

Kinetochore microtubules shorten

Polar microtubules lengthen

Centromeres split
(46 single chormosomes, 2N) (46 single chormosomes, 2N)

TELOPHASE

Nucleolus reappears

Chromosomes decondense

Nuclear envelope reforms

Polar microtubules lengthen

CYTOKINESIS

Contractile ring

Cleavage furrow
(46 single chormosomes, 2N) (46 single chormosomes, 2N)

FIGURE 9-1. Diagram of the stages of the M (mitosis) phase. Only one pair of homologous chromosomes (i.e., chromosome 18) is shown (white = maternal origin and black = paternal origin) for simplicity's sake.

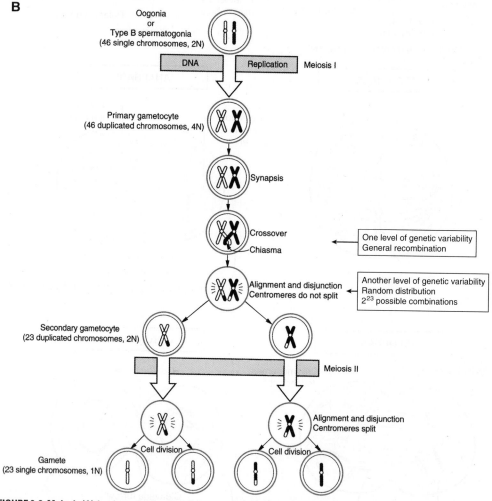

FIGURE 9-2. Meiosis (A) A schematic diagram of chromosome 18 shown in its "single chromosome" state and "duplicated chromosome" state that is formed by DNA replication during meiosis I. It is important to understand that both the "single chromosome" state and "duplicated chromosome" state will be counted as one chromosome 18. As long as the additional DNA in the "duplicated chromosome" is bound at the centromere, the structure will be counted as one chromosome 18 even though it has twice the amount of DNA. The "duplicated chromosome" is often referred to as consisting of two sister chromatids (chromatid 1 and chromatid 2). **(B)** Schematic representation of meiosis I and meiosis II, emphasizing the changes in chromosome number and amount of DNA that occur during gametogenesis. Only one pair of homologous chromosomes (i.e., chromosome 18) is shown (white = maternal origin and black = paternal origin) for simplicity sake. The point at which DNA crosses over is called the chiasma. Segments of DNA are exchanged thereby introducing genetic variability to the gametes. In addition, various cell types along with their appropriate designation of number of chromosomes and amount of DNA is shown.

t a b l e 9-1 Comparison of Meiosis and Mitosis

Meiosis	Mitosis
Occurs only in the testis and ovary	Occurs in a wide variety of tissues and organs
Produces haploid (23,1N) gametes (sperm and secondary oocyte)	Produces diploid (46, 2N) somatic daughter cells
Involves two cell divisions and one round of DNA replication	Involves one cell division and one round of DNA replication
Stages of Meiosis	**Stages of Mitosis**
Meiosis I	**Interphase**
Prophase	G_0 Phase
Leptotene (long, thin DNA strands)	G_1 Phase
Zygotene (synapsis occurs; synaptonemal complex)	←**G_1 Checkpoint**
Pachytene (crossover occurs: short. thick DNA strands)	S Phase
Diplotene (chromosomes separate except at centromere)	G_2 Phase
Prometaphase	←**G_2 Checkpoint**
Metaphase	
Anaphase	**Mitosis Phase**
Telophase	Prophase
	Prometaphase
Meiosis II (essentially identical to mitosis)	Metaphase
Prophase	Anaphase
Prometaphase	Telophase
Metaphase	
Anaphase	
Telophase	
Male: Prophase of meiosis I lasts ~22 days and completes meiosis II in a few hours	Interphase lasts ~15 hours M phase lasts ~1 hour
Female: Prophase of meiosis I lasts ~12 (puberty) – 50 years (menopause) and completes meiosis II when fertilization occurs	
Pairing of homologous chromosomes occurs	No pairing of homologous chromosomes
Genetic recombination occurs (exchange of large segments of maternal and paternal DNA via crossover during meiosis I)	Genetic recombination does not occur
Maternal and paternal homologous chromosomes are randomly distributed among the gametes to ensure genetic variability	Maternal and paternal homologous chromosomes are faithfully distributed among the daughter cells to ensure genetic similarity
Gametes are genetically different	Daughter cells are genetically identical

Review Test

1. The X and Y chromosomes pair in meiosis at the pseudoautosomal regions. A nondisjunction of the X and Y chromosomes in a male during meiosis I would produce which of the following combinations of gametes?

(A) one sperm with two X's, and three sperm with Y's

(B) two sperm with two X's and two sperm with two Y's

(C) one sperm with no X's, and three sperm with an X and a Y

(D) a sperm with two X's, a sperm with two Y's, and two sperm with no sex chromosomes

2. Which of the following describes the main difference between meiosis and mitosis?

(A) homologous chromosomes pair during meiosis

(B) the number of chromosomes is reduced by half during mitosis

(C) after meiosis is complete, there are 46 chromosomes in each cell

(D) after mitosis is complete there are 23 chromosomes in each cell.

3. Crossing over and random segregation produce much of the genetic variation in human populations. These events occur during which of the following?

(A) mitosis
(B) meiosis
(C) fertilization
(D) transcription

4. Tetraploid cells are the result of the failure of which one of the following processes?

(A) anaphase of mitosis
(B) S (synthesis) phase of the cell cycle
(C) cytokinesis of mitosis
(D) G_1 phase of the cell cycle

5. The "reduction division" in which the number of chromosomes in a germ cell is reduced from 46 to 23 chromosomes occurs during which of the following?

(A) mitosis
(B) meiosis I
(C) meiosis II
(D) synapsis

6. At the completion of oogenesis, how mature oocytes are formed from each primary oocyte?

(A) one
(B) two
(C) three
(D) four

7. Which one of the following has a haploid number of chromosomes?

(A) primary spermatocyte
(B) secondary spermatocyte
(C) spermatogonia
(D) oogonia

8. Karyotype analysis can be conducted on cells that have entered which one of the following stages of cell division?

(A) meiosis I
(B) meiosis II
(C) metaphase of mitosis
(D) anaphase of mitosis

9. One of the two places in the cell cycle where a response to DNA damage occurs is which one of the following?

(A) G_0 phase
(B) metaphase
(C) m (synthesis) phase
(D) G_2 checkpoint

Answers and Explanations

1. **The answer is (D).** During meiosis I, the nondisjunction of the paired and doubled X and Y would cause them to go into one of the daughter cells with no X or Y chromosomes going to the other daughter cell. During meiosis II, the doubled X chromosome and the doubled Y chromosome would go to separate daughter cells and the cell with no sex chromosomes would give rise to two daughter cells with no sex chromosomes. Because the X and Y chromosomes are doubled, the daughter cell receiving the X chromosomes will have two copies and the daughter cell receiving the Y chromosomes will have two copies.

2. **The answer is (A).** Homologous chromosomes pair during meiosis but not during mitosis. The number of chromosomes is reduced by half, from 46 to 23 during meiosis and the daughter cells are genetically different, but during mitosis, the chromosome number of 46 is maintained and the daughter cells are genetically identical.

3. **The answer is (B).** Crossing-over and random segregation of the maternal and paternal chromosomes occur during meiosis.

4. **The answer is (C).** The 46 doubled chromosomes separate during anaphase, resulting in 92 chromosomes, and if cytokinesis (cell division) fails, one cell with a tetraploid number of chromosomes, 92, is the result instead of two daughter cells with the normal diploid number of 46 in each.

5. **The answer is (B).** During meiosis I, the 23 paired, doubled homologs randomly separate, resulting in two daughter cells with 23 chromosomes each.

6. **The answer is (A).** Only one mature oocyte results from each primary oocyte. At each of the two cell divisions in meiosis, a polar body is formed, which usually degenerates.

7. **The answer is (B).** A secondary spermatocyte results from meiosis I, the reduction division, in a primary spermatocyte and the chromosome number is reduced from 46 to 23.

8. **The answer is (C).** Prometaphase and metaphase of mitosis is when the chromosomes are condensed enough to visualize for cytogenetic analysis.

9. **The answer is (D).** The other time in the cell cycle when there is a response to DNA damage is at the G_1 checkpoint before the S (synthesis) phase begins.

10 Chromosomal Morphology Methods

I. STUDYING HUMAN CHROMOSOMES

- Mitotic chromosomes are fairly easy to study because they can be observed in any cell undergoing mitosis.
- Meiotic chromosomes are much more difficult to study because they can be observed only in ovarian or testicular samples. In the female, meiosis is especially difficult because meiosis occurs during fetal development. In the male, meiotic chromosomes can be studied only in a testicular biopsy of an adult male.
- Any tissue that can be grown in culture can be used for **karyotype analysis,** but only certain tissue samples are suitable for some kinds of studies. For example, chorionic villi or amniocytes from amniotic fluid are used for prenatal studies; bone marrow is usually the most appropriate tissue for leukemia studies; skin or placenta is used for miscarriage studies; and blood for patients with dysmorphic features, unexplained mental retardation, or any other suspected genetic conditions.
- Whatever the tissue used, the cells must be grown in tissue culture for some period of time until optimal growth occurs. Blood cells must have a mitogen added to the culture media to stimulate the **mitosis** of lympocytes, but other tissues can be grown without such stimulation.
- Once a tissue has reached its optimal time for a harvest, **colchicine** (Colcemid) is added to the media, which arrests the cells in **metaphase.**
- The cells are then concentrated, treated with a hypotonic solution, which aids in the spreading of the chromosomes, and finally fixed with an acetic acid/methanol solution.
- The cell preparation is then dropped onto microscope slides and stained by a variety of methods (see below).
- It is often preferable to use **prometaphase** chromosomes in cytogenetic analysis as they are less condensed and therefore show more detail. In cytogenetic analysis, separated prometaphase or metaphase chromosomes are identified and photographed or digitized.
- The chromosomes in the photograph of the metaphase are then cut out and arranged in a standard pattern called the **karyotype,** or in the case of digital images, arranged into a karyotype with the assistance of a computer.

II. STAINING OF CHROMOSOMES

Metaphase or prometaphase chromosomes may be prepared for karyotype analysis and then stained by various techniques. In addition, one of the great advantages of some staining techniques is that metaphase or prometaphase chromosomes are not required.

A. **Chromosome Banding.** Chromosome banding techniques are based on denaturation and/or enzymatic digestion of DNA, followed by incorporation of a DNA-binding dye. This results in chromosomes staining as a series of dark and light bands.

   1. **G-Banding.** G-banding uses trypsin denaturation before staining with the Giemsa dye and is now the standard analytical method in cytogenetics.

      a. Giemsa staining produces a unique pattern of **dark bands (Giemsa positive; G bands)** which consist of heterochromatin, replicate in the late S phase, are rich in A-T bases, and contain few genes.

      b. Giemsa staining also produces a unique pattern of **light bands (Giemsa negative; R bands)** which consist of euchromatin, replicate in the early S phase, are rich in G-C bases, and contain many genes.

   2. **R-Banding.** R-banding uses the Giemsa dye (as above) to visualize **light bands (Giemsa negative; R bands)** which are essentially the *r*everse of the G-banding pattern. R-banding can also be visualized by G-C specific dyes (e.g., chromomycin A_3, oligomycin, or mithramycin).

   3. **Q-Banding.** Q-banding uses the fluorochrome *q*uinacrine (binds preferentially to A-T bases) to visualize **Q bands** which are essentially the same as G bands.

   4. **T-Banding.** T-banding uses severe heat denaturation prior to Giemsa staining or a combination of dyes and fluorochromes to visualize **T bands,** which are a subset of R bands, located at the *t*elomeres.

   5. **C-Banding.** C-banding uses barium hydroxide denaturation prior to Giemsa staining to visualize **C bands,** which are constitutive heterochromatin, located mainly at the centromere.

B. **Fluorescence *in situ* Hybridization (FISH).**

   ■ The FISH technique is based on the ability of single stranded DNA (i.e., a DNA probe) to hybridize (bind or anneal) to its complementary target sequence on a unique DNA sequence that one is interested in localizing on the chromosome.

   ■ Once this unique DNA sequence is known, a fluorescent DNA probe can be constructed.

   ■ The fluorescent DNA probe is allowed to hybridize with chromosomes prepared for karyotype analysis and thereby visualize the unique DNA sequence on specific chromosomes.

C. **Chromosome Painting.**

   ■ The chromosome painting technique is based on the construction of fluorescent DNA probes to a wide variety of different DNA fragments from a single chromosome.

   ■ The fluorescent DNA probes are allowed to hybridize with chromosomes prepared for karyotype analysis and thereby visualize many different loci spanning one whole chromosome (i.e., a chromosome paint). Essentially, one whole particular chromosome will fluoresce.

D. **Spectral Karyotyping or 24 Color Chromosome Painting.**

   ■ The spectral karyotyping technique is based on chromosome painting whereby DNA probes for all 24 chromosomes are labeled with five different fluorochromes so that each of the 24 chromosomes will have a different ratio of fluorochromes.

   ■ The different fluorochrome ratios cannot be detected by the naked eye but computer software can analyze the different ratios and assign a pseudocolor for each ratio.

   ■ This allows all 24 chromosomes to be painted with a different color. Essentially, all 24 chromosomes will be painted a different color.

   ■ The homologs of each chromosome will be painted the same color, but the X and Y chromosomes will be different colors, so 24 different colors are required.

E. **Comparative Genome Hybridization (CGH).**

   ■ The CGH technique is based on the competitive hybridization of two fluorescent DNA probes; one DNA probe from a normal cell labeled with a red fluorochrome and the other DNA probe from a tumor cell labeled with a green fluorochrome.

- The fluorescent DNA probes are mixed together and allowed to hybridize with chromosomes prepared for karyotype analysis.
- The ratio of red to green signal is plotted along the length of each chromosome as a distribution line.
- The red/green ratio should be 1:1 unless the tumor DNA is missing some of the chromosomal regions present in normal DNA (more red fluorochrome and the distribution line shifts to the left) or the tumor DNA has more of some chromosomal regions than present in normal DNA (more green fluorochrome and the distribution line shifts to the right).

III. CHROMOSOME MORPHOLOGY

A. The appearance of chromosomal DNA can vary considerably in a normal resting cell (e.g., degree of packaging, euchromatin, heterochromatin) and a dividing cell (e.g., mitosis and meiosis). It is important to note that the pictures of chromosomes seen in karyotype analysis are chromosomal DNA at a particular point in time i.e., arrested at metaphase (or prometaphase) of mitosis.

B. Early metaphase karyograms showed chromosomes as **X**-shaped because the chromosomes were at a point in mitosis when the protein **cohesin** no longer bound the sister chromatids together but the centromeres had not yet separated.

C. Modern metaphase karyograms show chromosomes as **I**–shaped because the chromosomes are at a point in mitosis when the protein cohesion still binds the sister chromatids together and the centromeres are not separated. In addition, many modern karyograms are prometaphase karyograms where the chromosomes are **I**-shaped.

IV. CHROMOSOME NOMENCLATURE

A. A chromosome consists of two characteristic parts called **arms**. The short arm is called the **p (petit) arm** and the long arm is called the **q (queue) arm.**

B. The arms of G-banded and R-banded chromosomes can be subdivided into **regions** (counting outwards from the centromere), **subregions (bands), sub-bands** (noted by the addition of a decimal point), and **sub-sub bands.**

C. For example, 6p21.34 is read as: the short arm of chromosome 6, region 2, subregion (band) 1, sub-band 3, and sub-sub band 4. This is *not* read as: the short arm of chromosome 6, twenty-one point thirty-four.

D. In addition, locations on an arm can be referred to in anatomical terms: **proximal** is closer to the centromere and **distal** is farther from the centromere.

E. The chromosome banding patterns of human G-banded chromosomes have been standardized and are represented diagrammatically in an idiogram.

F. A **metacentric chromosome** refers to a chromosome where the centromere is close to the midpoint, thereby dividing the chromosome into roughly equal length arms.

G. A **submetacentric chromosome** refers to a chromosome where the centromere is far away from the midpoint so that a p arm and q arm can be distinguished.

H. A **telocentric chromosome** refers to a chromosome where the centromere is at the very end of the chromosome so that only the q arm is described.

I. An **acrocentric chromosome** refers to a chromosome where the centromere is near the end of the chromosome, so that the p arm is very short (just discernible).

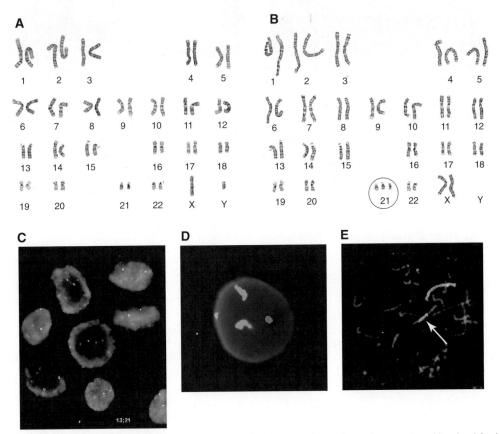

FIGURE 10-1. Karyotypes and chromosomal morphology. (A) G-banding of metaphase chromosomes with only minimal separation of the sister chromatids are shown arranged in a karyotype. Chromosomes 1 through 3 consist of the largest metacentric chromosomes. Chromosomes 4 and 5 are slightly smaller and submetacentric. Chromosomes 6 through12 are arranged in order of decreasing size with the centromere moving from a metacentric position to a submetacentric position. Chromosomes 13 through 15 are medium sized and acrocentric. Chromosomes 16 through18 are smaller and metacentric. Chromosomes 19 and 20 are even smaller and metacentric. Chromosomes 21 and 22 are the smallest chromosomes and acrocentric. The X chromosome is similar to chromosomes 6 through 12. The Y chromosome is similar to chromosomes 21 and 22. **(B) Karyotype of Down syndrome.** G-banding of metaphase chromosomes with only minimal separation of the sister chromatids are shown arranged in a karyotype. Note the three chromosomes 21 (circle). **(C) FISH for Down syndrome.** FISH using a probe for chromosome 21 (red dots) shows that each cell contains three red dots indicating trisomy 21. The green dots represent a control probe for chromosome 13. **(D) FISH for sex determination.** FISH using a probes for the X chromosome (green) and the Y chromosome (red) shows that a cell that contain one green dot and one red dot indicating the male sex. The two blue areas represent a control probe for chromosome 18. **(E) Chromosome painting.** Chromosome painting using paints for chromosome 4 (green) and chromosome 14 (red) shows a chromosomal rearrangement between chromosomes 4 and 14 (chromosome with green and red staining; arrow). *(continued)*

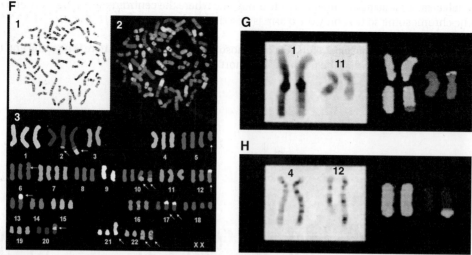

FIGURE 10-1. *(continued)* **(F) Spectral karyotyping of a chronic myelogenous leukemia cell line demonstrating a complex karyotype with several structural and numerical chromosome aberrations. (F1)** A metaphase cell showing the G-banding pattern. **(F2)** The same metaphase cell as in F1 showing the spectral display pattern. **(F3)** The same metaphase cell as in F1 and F2 arranged as a karyotype and stained with the spectral karyotyping colors. Arrows indicate structural chromosome aberrations involving two or more different chromosomes. **(G) Spectral karyotyping.** Spectral karyotyping using paints for chromosome 1 (yellow) and chromosome 11 (blue) shows a balanced reciprocal translocation between chromosomes 1 and 11, t(1q11p). A balance translocation means that there is no loss of any chromosomal segment during the translocation. This forms two derivative chromosomes each containing a segment of the other chromosome from the reciprocal exchange. **(H) Spectral karyotyping.** Spectral karyotyping using paints for chromosome 4 (blue) and chromosome 12 (red) shows an unbalanced reciprocal translocation between chromosomes 4 and 12, t(4q12q). An unbalanced translocation means that there is loss of a chromosomal segment during the translocation. In this case, the chromosomal segment 12 is lost.

Review Test

1. Which one of the following is a suitable specimen for cytogenetic analysis?

(A) placenta in formalin
(B) frozen (not cryopreserved) blood plasma
(C) frozen (not cryopreserved) amniotic fluid
(D) peripheral blood

2. Which one of the following is the appropriate specimen for cytogenetic analysis where the patient is a child with dysmorphic features and unexplained mental retardation?

(A) peripheral blood
(B) skin
(C) bone marrow
(D) cheek cells

3. A cytogenetics laboratory report states that a patient has a deletion of a chromosome distal to 5p15.31. Which of the following best describes what this means?

(A) There is a deletion of a portion of the long arm of chromosome 5 with the breakpoint at band p15.31.
(B) There is a deletion of a portion of the short arm of chromosome 5 with the breakpoint at band p15.31.
(C) There is a deletion of a portion of the long arm of chromosome 15 at band 5p31.
(D) There is a deletion of a portion of the short arm of chromosome 15 at band 5p31.

4. Which one of the following is often the preferred stage for more detailed cytogenetic analysis?

(A) meiotic prometaphase
(B) meiotic metaphase
(C) mitotic prometaphase
(D) mitotic metaphase

5. In a balanced reciprocal translocation in which two chromosomes exchange pieces, a breakpoint in which one of the following would be most likely to cause gene disruption and thus an abnormal phenotype?

(A) Giemsa negative G-band
(B) Giemsa positive G-band
(C) Giemsa negative R-band
(D) C-band

Answers and Explanations

1. **The answer is (D).** Tissues preserved in formalin and frozen tissues that have not been properly cryopreserved do not contain live cells, so they cannot be grown in culture.

2. **The answer is (A).** Peripheral blood is easily obtained and gives high quality cytogenetic preparations. A skin sample involves minor surgery. A bone marrow biopsy is painful and generally does not yield high quality cytogenetic preparations. Cheek cells are more appropriate for DNA studies because it would be difficult to obtain sufficient numbers of them for tissue culture and they would probably be too contaminated with bacteria to be grown successfully.

3. **The answer is (B).** The deletion is on the "p" or short arm of chromosome 5 at band 15.31.

4. **The answer is (C).** Meiotic chromosomes are not suitable for routine cytogenetic analysis. Metaphase chromosomes are suitable for cytogenetic analysis in general, but mitotic prometaphase chromosomes are more extended and allow for detailed, high-resolution cytogenetic analysis.

5. **The answer is (A).** The light Giemsa negative G-bands are GC-rich and contain more genes than the AT-rich G positive G-bands and the equivalent Giemsa negative R-bands. C-bands are heterochromatic and do not contain coding sequences.

I. NUMERICAL CHROMOSOMAL ABNORMALITIES

A. **Polyploidy** is the addition of an extra haploid set or sets of chromosomes (i.e., 23) to the normal diploid set of chromosomes (i.e., 46).
   1. **Triploidy** is a condition whereby cells contain **69 chromosomes.**
      a. Triploidy occurs as a result of either a **failure of meiosis in a germ cell** (e.g., fertilization of a diploid egg by a haploid sperm) or **dispermy** (two sperm that fertilize one egg).
      b. Triploidy results in spontaneous abortion of the conceptus or brief survival of the live-born infant after birth.
      c. **Partial hydatidiform mole.** A hydatidiform mole (complete or partial) represents an abnormal placenta characterized by marked enlargement of chorionic villi. A complete mole (no embryo present; see Chapter 1I-V-B) is distinguished from a partial mole (embryo present) by the amount of chorionic villous involvement. A partial mole occurs when ovum is fertilized by two sperm. This results in a **69, XXX or 69XXY karyotype** with one set of maternal chromosomes and two sets of paternal chromosomes.
   2. **Tetraploidy i**s a condition whereby cells contain **92 chromosomes.**
      a. Tetraploidy occurs as a result of **failure of the first cleavage division.**
      b. Tetraploidy almost always results in spontaneous abortion of the conceptus with survival to birth being an extremely rare occurrence.

B. **Aneuploidy** is the addition of one chromosome (**trisomy**), or loss of one chromosome (**monosomy**). Aneuploidy occurs as a result of **nondisjunction during meiosis.**
   1. **Trisomy 13 (Patau syndrome; 47,+13)**
      a. Trisomy 13 is a trisomic disorder caused by an extra chromosome 13.
      b. **Prevalence.** The prevalence of trisomy 13 is 1/20,000 live births. Live births usually die by ≈1 month of age. Most trisomy 13 conceptions spontaneously abort.
      c. **Clinical features include:** profound mental retardation, congenital heart defects, cleft lip and/or palate, omphalocele, scalp defects, and polydactyly.
   2. **Trisomy 18 (Edwards syndrome; 47,+18)**
      a. Trisomy 18 is a trisomic disorder caused by an extra chromosome 18.
      b. **Prevalence.** The prevalence of trisomy 18 is 1/5,000 live births. Live births usually die by ≈2 month of age. Most trisomy 18 conceptions spontaneously abort.
      c. **Clinical features include:** mental retardation, congenital heart defects, small facies and prominent occiput, overlapping fingers, cleft lip and/or palate, and rocker-bottom heels.
   3. **Trisomy 21 (Down syndrome; 47,+21)**
      a. Trisomy 21 is a trisomic disorder caused by an extra chromosome 21. Trisomy 21 is linked to a specific region on chromosome 21 called the **DSCR (Down syndrome critical region).** Trisomy 21 may also be caused by a specific type of translocation, called a **Robertsonian translocation** that occurs between acrocentric chromosomes.
      b. **Prevalence.** The prevalence of trisomy 21 is 1/2,000 conceptions for women <25 years of age, 1/300 conceptions for women ≈35 years of age, and 1/100 conceptions

in women ≈40 years of age. Trisomy 21 frequency increases with **advanced maternal age.**

   d. **Clinical features include:** moderate mental retardation (the leading cause of mental retardation), microcephaly, microphthalmia, colobomata, cataracts and glaucoma, flat nasal bridge, epicanthal folds, protruding tongue, simian crease in hand, increased nuchal skin folds, appearance of an "X" across the face when the baby cries, and congenital heart defects. Alzheimer neurofibrillary tangles and plaques are found in trisomy 21 patients after 30 years of age. A condition mimicking acute megakaryocytic leukemia (AMKL) frequently occurs in children with trisomy 21 and they are at increased risk for developing acute lymphoblastic leukemia (ALL).

4. Klinefelter syndrome (47, XXY)

   a. Klinefelter syndrome is a **trisomic** sex chromosome disorder caused by an extra X chromosome. The most common karyotype is 47,XXY but other karyotypes (e.g., 48,XXXY) and **mosaics** (47,XXY/ 46,XY) have been reported.
   b. Klinefelter syndrome is **found only in males** and is associated with **advanced paternal age.**
   c. **Prevalence.** The prevalence of Klinefelter syndrome is 1/1,000 live male births.
   d. **Clinical features include:** varicose veins, arterial and venous leg ulcer, scant body and pubic hair, male hypogonadism, sterility with fibrosus of seminiferous tubules, marked decrease in testosterone levels, elevated gonadotropin levels, gynecomastia, IQ slightly less than that of siblings, learning disabilities, antisocial behavior, delayed speech as a child, tall stature, and eunuchoid habitus.

5. Turner syndrome (Monosomy X; 45,X)

   a. Monosomy X is a **monosomic** sex chromosome disorder caused by a loss of part or all of the X chromosome. ≈66% of monosomy X females retain the maternal X chromosome and 33% retain the paternal X chromosome. ≈50% of monosomy X females are **mosaics** [e.g., 45,X/46,XX or 45,X/46, +i(Xq)].
   b. Monosomy X is the only monosomic disorder compatible with life and is **found only in females.**
   c. The **SHOX gene** (**sho**rt stature homeobox-containing gene on the **X** chromosome) which encodes for the **short stature homeobox protein** is most likely one of the genes that is deleted in Monosomy X and results in the short stature of these females.
   d. **Prevalence.** The prevalence of monosomy X is ≈1/2,000 live female births. There are ≈50,000 to 75,000 monosomy X females in the U.S. population, although true prevalence is difficult to calculate because monosomy X females with mild phenotypes remain undiagnosed. ≈3% of all female conceptions results in monosomy X making it the most common sex chromosome abnormality in female conceptions. However, most monosomy X female conceptions spontaneously abort.
   e. **Clinical features include:** short stature, low-set ears, ocular hypertelorism, ptosis, low posterior hairline, webbed neck due to a remnant of a fetal cystic hygroma, congenital hypoplasia of lymphatics causing peripheral edema of hands and feet, shield chest, pinpoint nipples, congenital heart defects, aortic coarctation, female hypogonadism, ovarian fibrous streaks (i.e., infertility), primary amenorrhea, and absence of secondary sex characteristics.

C. Mixoploidy. Mixoploidy is a condition where a person has two or more genetically different cell populations. If the genetically different cell populations arise from a single zygote, the condition is called **mosaicism.** If the genetically different cell populations arise from different zygotes, the condition is called **chimerism.**

1. Mosaicism

   - A person may become a mosaic by **postzygotic mutations** that can occur at any time during postzygotic life.
   - These postzygotic mutations are actually quite frequent in humans and produce genetically different cell populations (i.e., most of us are mosaics to a certain extent). However, these postzygotic mutations are not usually clinically significant.
   - If the postzygotic mutation produces a substantial clone of mutated cells, then a clinical consequence may occur.

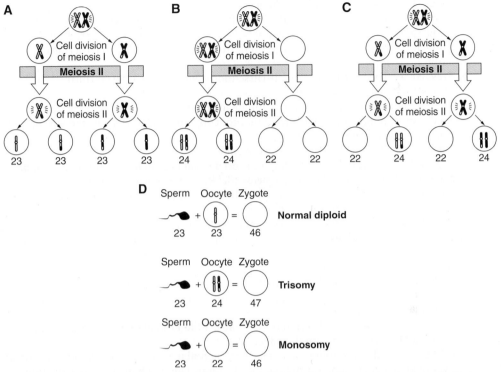

FIGURE 11-1. Meiosis and nondisjunction. (A) Normal meiotic divisions (I and II) producing gametes with 23 chromosomes. **(B)** Nondisjunction occurring in meiosis I producing gametes with 24 and 22 chromosomes. **(C)** Nondisjunction occurring in meiosis II producing gametes with 24 and 22 chromosomes. **(D)** Although nondisjunction may occur in either spermatogenesis or oogenesis, there is a higher frequency of nondisjunction in oogenesis. In this schematic, nondisjunction in oogenesis in depicted. If an abnormal oocyte (24 chromosomes) is fertilized by a normal sperm (23 chromosomes), a zygote with 47 chromosomes is produced (i.e., trisomy). If an abnormal oocyte (22 chromosomes) is fertilized by a normal sperm (23 chromosomes), a zygote with 45 chromosomes is produced (i.e., monosomy).

- The formation of a substantial clone of mutated cells can occur in two ways: the mutation results in an abnormal proliferation of cells (e.g., formation of cancer) or the mutation occurs in a progenitor cell during early embryonic life and forms a significant clone of mutated cells.

- A postzygotic mutation may also cause a clinical consequence if the mutation occurs in the germ-line cells of a parent (called **germinal or gonadal mosaicism**). For example, if a postzygotic mutation occurs in male spermatogenic cells, then the man may harbor a large clone of mutant sperm without any clinical consequence (i.e., the man is normal). However, if the mutant sperm from the normal male fertilizes a secondary oocyte, the infant may have a *de novo* inherited disease. This means that a normal couple without any history of inherited disease may have a child with a *de novo* inherited disease if one of the parents is a gonadal mosaic.

2. **Chimerism.** A person may become a chimera by the fusion of two genetically different zygotes to form a single embryo (i.e., the reverse of twinning) or by the limited colonization of one twin by cells from a genetically different (i.e., nonidentical; fraternal) co-twin.

II. STRUCTURAL CHROMOSOMAL ABNORMALITIES

A. **Deletions** are a loss of chromatin from a chromosome. There is much variability in the clinical presentations based on what particular genes and the number of genes that are deleted. Some of the more common deletion abnormalities are indicated below.

1. **Chromosome 4p deletion (Wolf-Hirschhorn syndrome; WHS)**
   a. WHS is caused by a deletion of the **Wolf-Hirschhorn critical region (WHCR)** on **chromosome 4p16.3** ≈75% of WHS individuals have a *de novo* deletion, 13% inherited an unbalanced chromosome rearrangement from a parent, and 12% have a ring chromosome 4.
   b. **Prevalence.** The prevalence of Wolf-Hirschhorn syndrome is 1/50,000 births, with a 2:1 female/male ratio.
   c. **Clinical features include:** prominent forehead and broad nasal root ("Greek warrior helmet"), short philtrum, down-turned mouth, congenital heart defects, growth retardation, and severe mental retardation.

2. **Chromosome 5p deletion (Cri du chat; cat cry syndrome)**
   a. Cri du chat is caused by a deletion of the **cri du chat critical region (CDCCR)** on **chromosome 5p15.2** and the **catlike critical region (CLCR)** on **chromosome 5p15.3**. ≈80% of cri du chat individuals have a *de novo* deletion. In ≈80% of the cases, the deletions occur on the paternal chromosome 5.
   b. **Prevalence.** The prevalence of cri du chat syndrome is 1/50,000 births.
   c. **Clinical features include:** round facies, a catlike cry, congenital heart defects, microcephaly, and mental retardation.

B. **Microdeletions** are a loss of chromatin from a chromosome that cannot be detected easily, even by high-resolution banding. FISH is the definitive test for detecting microdeletions.
   1. **Prader-Willi syndrome (PW)**
      a. PW is caused by a microdeletion of the **Prader-Willi critical region (PWCR)** on **chromosome 15q11.2-13** derived from the **father.**
      b. PW illustrates the phenomenon of **genomic imprinting** which is the differential expression of genes depending on the parent of origin. The mechanism of inactivation (or genomic imprinting) involves **DNA methylation of cytosine nucleotides** during gametogenesis resulting in transcriptional inactivation.
      c. The counterpart of PW is **Angelman syndrome.** Other examples that highlight the role of genomic imprinting include **complete hydatidiform moles** and **Beckwith-Wiedemann syndrome (BWS)** (see Chapter 1IV).
      d. The paternally inherited *SNRPN* **allele,** which encodes for a **small nuclear ribonucleoprotein-associated N protein** is most likely one of the genes that is deleted in PW and results in some of the clinical features of PW.
      d. **Prevalence.** The prevalence of PW is 1/10,000 to 25,000 births.
      e. **Clinical features include:** poor feeding and hypotonia at birth, but then followed by hyperphagia (insatiable appetite), hypogonadism, obesity, short stature, small hands and feet, behavior problems (rage, violence), and mild-to-moderate mental retardation.
   2. **Angelman syndrome (AS; happy puppet syndrome)**
      a. AS is caused by a microdeletion of the **AS/PWS region** on **chromosome 15q11.2-13** derived from the **mother.**
      b. AS is an example of **genomic imprinting** (see above). The counterpart of AS is **Prader-Willi syndrome.**
      c. The maternally inherited *UBE3A* **allele** which encodes for **ubiquitin-protein ligase E3A** is most likely one of the genes that is deleted in AS and results in many of the clinical features of AS. The loss of ubiquitin-protein ligase E3A disrupts the protein degradation pathway.
      d. **Prevalence.** The prevalence of AS is 1/12,000 to 20,000 births.
      e. **Clinical features include:** gait ataxia (stiff, jerky, unsteady, upheld arms), seizures, happy disposition with inappropriate laughter, severe mental retardation (only 5 to 10 word vocabulary), developmental delays are noted at ≈6 months, and age of onset ≈1 year of age.
   3. **22q11.2 Deletion syndrome (DS)**
      a. DS is caused by a microdeletion of the **DiGeorge chromosomal critical region (DGCR)** on **chromosome 22q11.2.** ≈90% of DS individuals have a *de novo* deletion.

b. The **TBX1 gene,** which encodes for **T-box transcription factor TBX10 protein** is most likely one of the genes that is deleted in DS and results in some of the clinical features of DS.

c. DS encompasses the phenotypes previously called **DiGeorge syndrome, velocardiofacial syndrome, conotruncal anomaly face syndrome, Opitz g/BBB syndrome,** and **Cayler cardiofacial syndrome.**

d. **Prevalence.** The prevalence of DS is 1/6,000 births in the U.S. population.

e. **Clinical features include:** facial anomalies resembling first arch syndrome (micrognathia, low-set ears) due to abnormal neural crest cell migration, cardiovascular anomalies due to abnormal neural crest cell migration during formation of the aorticopulmonary septum (e.g., Tetralogy of Fallot), velopharyngeal incompetence, cleft palate, immunodeficiency due to thymic hypoplasia, hypocalcemia due to parathyroid hypoplasia, and embryological formation of pharyngeal pouches 3 and 4 fail to differentiate into the thymus and parathyroid glands.

4. **Miller-Dieker syndrome (MD; agyria; lissencephaly)**

a. MD is caused by a microdeletion on **chromosome 17p13.3.**

b. The **LIS1 gene** (**lis**sencephaly) which encodes for the **LIS1 protein** is most likely one of the genes that is deleted in MD and results in some of the clinical features of MD. The LIS1 protein contains a coiled-coil domain and a tryptophan-aspartate repeat domain both of which interact with microtubules and multiprotein complexes within migrating neurons.

c. The **14-3-3ε gene,** which encodes for the **14-3-3ε protein** is another likely gene deleted in MD and results in some of the clinical features of MD. The 14-3-3ε protein phosphorylated serine and phosphorylated threonine domains both of which interact with microtubules and multiprotein complexes within migrating neurons.

d. **Prevalence.** The prevalence of MD is unknown.

e. **Clinical features include:** lissencephaly (smooth brain, i.e., no gyri), microcephaly, a high and furrowing forehead, death occurs early. Lissencephaly should not be mistakenly diagnosed in the case of premature infants whose brains have not yet developed an adult pattern of gyri (gyri begin to appear normally at about week 28).

5. **WAGR syndrome**

a. WAGR is caused by a microdeletion on **chromosome 11p13.** ≈90% of WAGR individuals have a *de novo* deletion.

b. The **WT1 gene** (**W**ilms **t**umor gene 1) which encodes for the **WT1 protein** (a zinc finger DNA-binding protein) is most likely one of the genes that is deleted in WAGR and results in the genitourinary clinical features of WAGR. WT1 protein is required for the normal embryological development of the genitourinary system. WT1 protein isoforms synergize with **SF-1** (steroidogenic factor-1) which is a nuclear receptor that regulates the transcription of a number of genes involved in reproduction, steroidogenesis, and male sexual development.

c. The **PAX6 gene** (**pa**ired bo**x**), which encodes for the **PAX6 protein** (a paired box transcription factor) is another likely gene that is deleted in WAGR and results in the aniridia and mental retardation clinical features of WAGR.

d. **Prevalence.** The prevalence of WAGR syndrome is unknown. However, the prevalence of Wilms tumor is 1/125,000 in the U.S. population.

e. **Clinical features include: W**ilms tumor, **a**niridia (absence of the iris), **g**enitourinary abnormalities (e.g., gonadoblastoma), and mental **r**etardation. Wilms tumor is the **most common renal malignancy of childhood,** which usually presents between 1 to 3 years of age. WT presents as a large, solitary, well-circumscribed mass that on cut section is soft, homogeneous, and tan–gray in color. WT is interesting histologically in that this tumor tends to recapitulate different stages of embryological formation of the kidney so that three classic histological areas are described: a stromal area, a blastemal area of tightly packed embryonic cells, and a tubular area. In 95% of the cases, the WT tumor is sporadic and unilateral.

6. **Williams syndrome (WS)**

a. **WS** is caused by a microdeletion of the **Williams-Beuren syndrome critical region (WBSCR)** on **chromosome 7q11.23.** ≈90% of WS individuals have a *de novo* deletion.

b. The **ELN gene** (elastin) which encodes for the **elastin protein** is most likely one of the genes that is deleted in WS and results in some of the clinical features of WS.

c. The **LIMK1 gene,** which encodes for a **brain-expressed lim kinase 1 protein** is another likely gene that is deleted in WS and results in some of the clinical features of WS.

d. Prevalence. The prevalence of WS is 1/7,500 in a Norway population.

e. Clinical features include: facial dysmorphology (e.g., prominent lips, wide mouth, periorbital fullness of subcutaneous tissues, short palpebral tissues, short upturned nose, long philtrum), cardiovascular disease (e.g., elastin arteriopathy, supravalvular aortic stenosis, pulmonic valvular stenosis, hypertension, septal defects), endocrine abnormalities (e.g., hypercalcemia, hypercalciuria, hypothyroidism, early puberty), prenatal growth deficiency, failure to thrive in infancy, connective tissue abnormalities (e.g., hoarse voice, hernias, rectal prolapse, joint and skin laxity), and mild mental deficiency with uneven cognitive disabilities.

C. Translocations result from breakage and exchange of segments between chromosomes.

1. Robertsonian translocation (RT)

- An RT is caused by translocations between the long arms (q) of acrocentric (satellite) chromosomes where the breakpoint is near the centromere. The short arms (p) of these chromosomes are generally lost.

- Carriers of an RT are **clinically normal** because the short arms, which are lost, contain only inert DNA and some rRNA (ribosomal RNA) genes, which occur in multiple copies on other chromosomes.

- One of the most common translocations found in humans is the **Robertsonian translocation t(14q21q).**

- The clinical issue in the Robertsonian translocation t(14q21q) occurs when the carriers produce gametes by meiosis and reproduce. Depending on how the chromosomes segregate during meiosis, conception can produce offspring with translocation trisomy 21 (live birth), translocation trisomy 14 (early miscarriage), monosomy 14 or 21 (early miscarriage), a normal chromosome complement (live birth), or a t(14q21q) carrier (live birth).

- A couple where one member is a t(14q21q) carrier may have a baby with translocation trisomy 21 (Down syndrome) or recurrent miscarriages.

2. Reciprocal translocation (RC)

- An RC is caused by the exchange of segments between two chromosomes, which forms two derivative (der) chromosomes each containing a segment of the other chromosome from the reciprocal exchange.

- **b.** One of the most common inherited reciprocal translocations found in humans is the **t(11;22)(q23.3;q11.2).**

- The translocation heterozygote, or carrier, would be at risk of having a child with abnormalities due to passing on only one of the derivative chromosomes. That would result in a child who would be partially trisomic for one of the participant chromosomes and partially monosomic for the other.

3. Acute promyelocytic leukemia (APL) t(15;17)(q22;q21)

a. APL t(15;17)(q22;q21) is caused by a reciprocal translocation between chromosomes 15 and 17 with breakpoints at bands q22 and q21, respectively.

b. This results in a fusion of the **promyelocyte gene (PML gene)** on 15q22 with the **retinoic acid receptor gene (RARα gene)** on 17q21, thereby forming the **PML/RARα oncogene.**

c. The **PML/RARα oncoprotein** (a transcription factor) blocks the differentiation of promyelocytes to mature granulocytes such that there is continued proliferation of promyelocytes.

d. Clinical features include: pancytopenia (i.e., anemia, neutropenia, and thrombocytopenia), including weakness and easy fatigue, infections of variable severity, and/or hemorrhagic findings (e.g., gingival bleeding, ecchymoses, epistaxis, or menorrhagia), and bleeding secondary to disseminated intravascular coagulation. A rapid cytogenetic

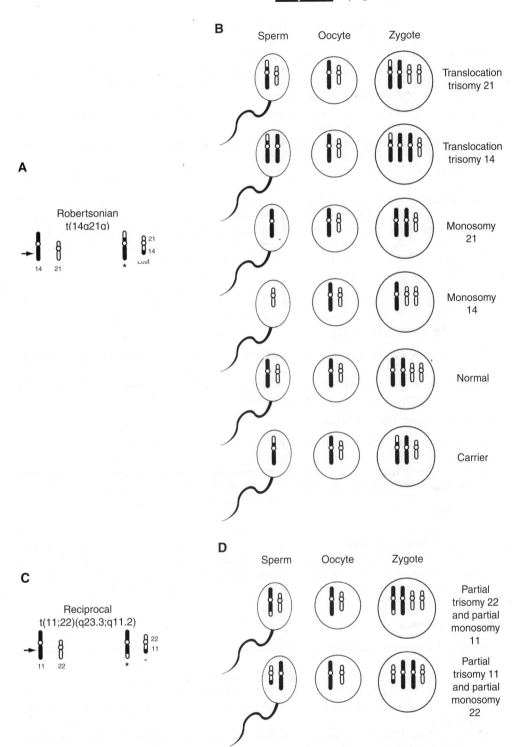

FIGURE 11-2. Translocations. (A) Robertsonian t(14q21q). This is one of the most common Robertsonian translocations found in humans. **(B)** Diagram shows the six conditions that may result depending on how chromosomes 14 and 21 segregate during meiosis when the carrier of the Robertsonian translocation is the male. * = robertsonian translocation chromosome. **(C) Reciprocal translocation t(11;22)(q23.3;q11.2).** This is one of the most common reciprocal translocations found in humans. **(D)** Diagram shows the two conditions that may result depending on how chromosomes 11 and 22 segregate during meiosis when the carrier of the reciprocal translocation is the male. * = reciprocal translocation chromosome

diagnosis of this leukemia is essential for patient management because these patients are at an extremely high risk for stroke.

4. Chronic myeloid leukemia (CML) t(9;22)(q34;q11.2)

a. CML t(9;22)(q34;q11.2) is caused by a reciprocal translocation between chromosomes 9 and 22 with breakpoints at q34 and q11.2 respectively. The resulting der(22) is referred to as the **Philadelphia chromosome.**

b. This results in a fusion of the *ABL* **gene** on 9q34 with the *BCR* **gene** on 22q11.1, thereby forming the *ABL/BCR* oncogene.

c. The **ABL/BCR oncoprotein** (a tyrosine kinase) has enhanced tyrosine kinase activity that transforms hematopoietic precursor cells.

d. Prevalence. The prevalence of CML is 1/100,000 per year with a slight male predominance.

d. Clinical features include: systemic symptoms (e.g., fatigue, malaise, weight loss, excessive sweating), abdominal fullness, bleeding episodes due to platelet dysfunction, abdominal pain may include left upper quadrant pain, early satiety due to the enlarged spleen, tenderness over the lower sternum due to an expanding bone marrow, and the uncontrolled production of maturing granulocytes, predominantly neutrophils, but also eosinophils and basophils.

D. Isochromosomes occur when the centromere divides transversely (instead of longitudinally) such that one of the chromosome arms is duplicated and the other arm is lost.

1. Isochromosome Xq [46,+i (Xq)]

- Isochromosome Xq is caused by a duplication of the q arm and loss of p arm of chromosome X.
- Isochromosome Xq is found in 20% of females with **Turner syndrome,** usually as a mosaic cell line along with a 45,X cell line [i.e.,45,X/46, +i(Xq)].

2. Isochromosome 12p [47,+i (12p)]

- The occurrence of isochromosomes within any of the autosomes is generally a lethal situation although isochromosomes for small segments do allow for survival to term.
- Isochromosome 12p is associated with **testicular germ cell tumors.** The *CCND2* **gene** located on **chromosome 12p13** encodes for *cyclin D2,* which regulates the cell cycle at the G1 checkpoint. Overexpression of cyclin D2 has been demonstrated in a variety of testicular germ cell tumors.
- Isochromosome 12p is also associated with a rare polydysmorphic syndrome called **Pallister-Killian syndrome.** Clinical features include: mental retardation, loss of muscle tone, streaks of skin with hypopigmentation, high forehead, coarse facial features, wide space between the eyes, broad nasal bridge, highly arched palate, fold of skin over the inner corner of the eyes, large ears, joint contractures, and cognitive delays.

E. Ring Chromosomes

- Ring chromosomes are formed when breaks occur somewhere on either side of the centromere.
- The newly created fragments (and thus the genes on them) are lost and the remaining pieces of the short and long arms join with each, forming a ring.
- Ring chromosomes are unstable and tend to be lost during mitosis, creating a mosaic cell line.
- A ring chromosome X is found in ≈15% of individuals with Turner syndrome, usually as a mosaic cell line with a 45,X cell line.

F. Inversions

- Inversions are the reversal of the order of DNA between two breaks in a chromosome.
- **Pericentric inversion** breakpoints occur on both sides of the centromere.
- **Paracentric inversion** breakpoints occur on the same side of the centromere.

- Carriers of inversions are usually normal. The diagnosis of an inversion is generally a coincidental finding during prenatal testing or the repeated occurrence of spontaneous abortions or stillbirths.
- The risk for an inversion carrier to have a child with an abnormality or to have reproductive loss is due to crossing-over in the inversion loop that forms during meiosis as the normal and inverted chromosomes pair.
- When the chromosomes separate, duplications and deletions of chromosomal material occur.

G. Chromosome breakage is caused by breaks in chromosomes due to sunlight (or ultraviolet) irradiation, ionizing irradiation, DNA crosslinking agents, or DNA damaging agents. These insults may cause **depurination of DNA, deamination of cytosine to uracil,** or **pyrimidine dimerization**, which must be repaired by DNA repair enzymes.

1. **Xeroderma pigmentosum (XP)**
   a. XP is an autosomal recessive genetic disorder caused by mutations in **nucleotide excision repair enzymes,** which results in the inability to remove pyrimidine dimers and individuals who are hypersensitive to **sunlight (UV radiation).**
   b. The **XPA gene** and the **XPC gene** are two of the genes involved in the cause of XP. *XPA* gene located on chromosome 9q22.3 encodes for a DNA repair enzyme. The *XPC* gene located on chromosome 3p25 also encodes for a DNA repair enzyme
   c. **Prevalence.** The prevalence of XP is 1/250,000 in the U.S. population.
   d. **Clinical features include:** sunlight (UV radiation) hypersensitivity with sunburnlike reaction, severe skin lesions around the eyes and eyelids, and malignant skin cancers (basal and squamous cell carcinomas and melanomas) whereby most individuals die by 30 years of age.

2. **Ataxia-telangiectasia (AT)**
   a. AT is an autosomal recessive genetic disorder caused by mutations in **DNA recombination repair enzymes** on chromosome 11q22-q23, which results in individuals who are hypersensitive to **ionizing radiation.**
   b. The **ATM gene (AT mutated)** is one of the genes involved in the cause of AT. The *ATM* gene located on chromosome 11q22 encodes for a protein where one region resembles a **PI-3 kinase** (phosphatidylinositol-3 kinase) and another region resembles a **DNA repair enzyme/cell cycle checkpoint protein.**
   c. **Prevalence.** The prevalence of AT is 1/20,000 to 100,000 in the U.S. population.
   d. **Clinical features include:** ionizing radiation hypersensitivity, cerebellar ataxia with depletion of Purkinje cells, progressive nystagmus, slurred speech, oculocutaneous telangiectasia initially in the bulbar conjunctiva followed by ear, eyelid, cheeks, and neck, immunodeficiency, and death in the second decade of life. A high frequency of structural rearrangements of chromosomes 7 and 14 is the cytogenetic observation with this disease.

3. **Fanconi anemia (FA)**
   a. FA is an autosomal recessive genetic disorder caused by mutations in **DNA recombination repair,** which results in individuals who are hypersensitive to **DNA crosslinking agents.**
   b. The **FA-A gene** (involved in 65% of FA cases) is one of the genes involved in the cause of FA. The *FA-A* gene located on chromosome 16q24 encodes for a protein that normalizes cell growth, corrects sensitivity to chromosomal breakage in the presence of mitomycin C, and generally promotes genomic stability.
   c. **Prevalence.** The prevalence of FA is 1/32,000 in the Ashkenazi Jewish population.
   d. **Clinical features include:** DNA crosslinking agent hypersensitivity, short stature, hypopigmented spots, café-au-lait spots, hypogonadism microcephaly, hypoplastic or aplastic thumbs, renal malformation including unilateral aplasia or horseshoe kidney, acute leukemia, progressive aplastic anemia, head and neck tumors, medulloblastoma, and is the most common form of congenital aplastic anemia.

4. **Bloom syndrome (BS)**
   a. BS is an autosomal recessive genetic disorder caused by mutations **DNA repair enzymes** on chromosome 15q26 which results in individuals who are hypersensitive to **DNA-damaging agents.**

b. The **BLM gene** is one of the genes involved in the cause of BS. The *BLM* gene located on chromosome 15q26 encodes for **RecQ helicase,** which unwinds the DNA double helix during repair and replication.

c. Prevalence. The prevalence of BS is high in the Ashkenazi Jewish population.

d. Clinical features include: hypersensitivity to DNA-damaging agents, long, narrow face, erythema with telangiectasias in butterfly distribution over the nose and cheeks, high-pitched voice, small stature, small mandible, protuberant ears, absence of upper lateral incisors, well-demarcated patches of hypopigmentation and hyperpigmentation, immunodeficiency with decreased IgA, IgM, and IgG levels, and predisposition to several types of cancers.

5. Hereditary nonpolyposis colorectal cancer (HNPCC)

a. HNPCC is an autosomal dominant genetic disorder caused by mutations in **DNA mismatch repair enzymes, which** results in the inability to remove single nucleotide mismatches or loops that occur in microsatellite repeat areas.

b. The four genes involved in the cause of HNPCC include:

  i. *MLH*1 **gene** located on chromosome 3p21.3 which encodes for DNA mismatch repair proteinMlh1.

  ii. *MSH2* **gene** located on chromosome 2p22-p21, which encodes for DNA mismatch repair protein Msh2

  iii. *MSH6* **gene** located on chromosome 2p16 which encodes for DNA mismatch repair protein Msh6

  iv. *PMS2* **gene** located on chromosome 7p22 which encodes for PMS1 protein homolog 2.

c. These genes are the human homologues to the *Escherichia coli* **mutS gene** and **mutI gene** that code for DNA mismatch repair enzymes.

d. Prevalence. HNPCC accounts for 1% to 3% of colon cancers and ≈1% of endometrial cancers.

e. Clinical features include: onset of colorectal cancer at a young age, high frequency of carcinomas proximal to the splenic flexure, multiple synchronous or metachronous colorectal cancers, and presence of extracolonic cancers (e.g., endometrial and ovarian cancer, adenocarcinomas of the stomach, small intestine, and hepatobiliary tract), and accounts for 3% to 5% of all colorectal cancers.

III. SUMMARY TABLE OF CYTOGENETIC DISORDERS (Table 11-1)

IV. SELECTED PHOTOGRAPHS OF CYTOGENETIC DISORDERS (Figures 11-3, 11-4, 11-5, 11-6)

table **11-1** Summary Table of Cytogenetic Disorders

Cytogenetic Disorder	Chromosomal Defect	Clinical Features
Numerical Chromosomal Abnormalities (Aneuploidy)		
Trisomy 13 (Patau syndrome; 47,+13)	Aneuploidy; 13	Profound mental retardation, congenital heart defects, cleft lip and/or palate, omphalocele, scalp defects, and polydactyly
Trisomy 18 (Edwards syndrome; 47,+18)	Aneuploidy; 18	Mental retardation, congenital heart defects, small facies and prominent occiput, overlapping fingers, cleft lip and/or palate, and rocker-bottom heels
Trisomy 21 (Down syndrome; 47,+21)	Aneuploidy; 21 DSCR	Moderate mental retardation (the leading cause of mental retardation), microcephaly, microphthalmia, colobomata, cataracts and glaucoma, flat nasal bridge, epicanthal folds, protruding tongue, simian crease in hand, increased nuchal skin folds, appearance of an "X" across the face when the baby cries, and congenital heart defects. Alzheimer neurofibrillary tangles and plaques are found in Down syndrome patients after 30 years of age. A condition mimicking acute megakaryocytic leukemia (AMKL) frequently occurs in children with Down syndrome and they are at increased risk for developing acute lymphoblastic leukemia (ALL)
Trisomy 47, XXY (Klinefelter syndrome; 47,XXY; 48,XXXY; 47,XXY/46,XY)	Aneuploidy; extra X	Varicose veins, arterial and venous leg ulcer, scant body and pubic hair, male hypogonadism, sterility with fibrosus of seminiferous tubules, marked decrease in testosterone levels, elevated gonadotropin levels, gynecomastia, IQ slightly less than that of siblings, learning disabilities, antisocial behavior, delayed speech as a child, tall stature and eunuchoid habitus, **found only in males**
Monosomy X (Turner syndrome; 45,X; 45,X/46,XX; 45,X/46, +iXq)	Aneuploidy; loss of X SHOX gene	Short stature, low-set ears, ocular hypertelorism, ptosis, low posterior hairline, webbed neck due to a remnant of a fetal cystic hygroma, congenital hypoplasia of lymphatics causing peripheral edema of hands and feet, shield chest, pinpoint nipples, congenital heart defects, aortic coarctation, female hypogonadism, ovarian fibrous streaks (i.e., infertility), primary amenorrhea, and absence of secondary sex characteristics, **found only in females**
Structural Chromosomal Abnormalities (Deletions/Microdeletions)		
Wolf-Hirschhorn syndrome	4p16.3 deletion WHCR	Prominent forehead and broad nasal root ("Greek warrior helmet"), short philtrum, down-turned mouth, congenital heart defects, growth retardation, and severe mental retardation
Cri du chat syndrome	5p15.2 deletion CDCCR CLCR	Round facies, a catlike cry, congenital heart defects, microcephaly, and mental retardation
Prader-Willi syndrome	Paternal 15q11.2-13 microdeletion; Imprinting SNRPN allele	Poor feeding and hypotonia at birth, but then followed by hyperphagia (insatiable appetite), hypogonadism, obesity, short stature, small hands and feet, behavior problems (rage, violence), and mild-to-moderate mental retardation
Angelman syndrome	Maternal 15q11.2-13 microdeletion; Imprinting UBE3A allele	Gait ataxia (stiff, jerky, unsteady, upheld arms), seizures, happy disposition with inappropriate laughter, severe mental retardation (only 5–10 word vocabulary), developmental delays are noted at ≈6 months, and age of onset is ≈1 year of age
22q11.2 Deletion syndrome (DiGeorge, Velocardiofacial, Conotruncal anomaly face, Opitz/BBB, Cayler cardiofacial)	22q11.2 microdeletion DGCR TBX1 gene	Facial anomalies resembling first arch syndrome (micrognathia, low-set ears) due to abnormal neural crest cell migration, cardiovascular anomalies due to abnormal neural crest cell migration during formation of the aorticopulmonary septum (e.g., tetralogy of Fallot), velopharyngeal incompetence, cleft palate, immunodeficiency due to thymic hypoplasia, hypocalcemia due to parathyroid hypoplasia, and embryological formation of pharyngeal pouches 3 and 4 fail to differentiate into the thymus and parathyroid glands

(continued)

t a b l e   **11-1**   *(continued)*

Cytogenetic Disorder	Chromosomal Defect	Clinical Features
Structural Chromosomal Abnormalities (Deletions/Microdeletions)		
Miller-Dieker syndrome	17p13.3 microdeletion *LIS1* gene *14-3-3ε* gene	Lissencephaly (smooth brain, i.e., no gyri), microcephaly, a high and furrowing forehead; death occurs early. Lissencephaly should not be mistakenly diagnosed in the case of premature infants whose brains have not yet developed an adult pattern of gyri (gyri begin to appear normally at about week 28)
WAGR syndrome	11p13 microdeletion *WT1* gene *PAX6* gene	**W**ilms tumor, **a**niridia (absence of the iris), **g**enitourinary abnormalities (e.g., gonadoblastoma), and mental **r**etardation. Wilms tumor is the **most common renal malignancy of childhood**, which usually presents between 1–3 years of age. WT presents as a large, solitary, well-circumscribed mass that on cut section is soft, homogeneous, and tan–gray in color. WT is interesting histologically in that this tumor tends to recapitulate different stages of embryological formation of the kidney so that three classic histological areas are described: a stromal area, a blastemal area of tightly packed embryonic cells, and a tubular area. In 95% of the cases, the WT tumor is sporadic and unilateral
Williams syndrome	7q11.23 microdeletion WBSCR *ELN* gene *LIMK1* gene	Facial dysmorphology (e.g., prominent lips, wide mouth, periorbital fullness of subcutaneous tissues, short palpebral tissues, short upturned nose, long philtrum), cardiovascular disease (e.g., elastin arteriopathy, supravalvular aortic stenosis, pulmonic valvular stenosis, hypertension, septal defects), endocrine abnormalities (e.g., hypercalcemia, hypercalciuria, hypothyroidism, early puberty), prenatal growth deficiency, failure to thrive in infancy, connective tissue abnormalities (e.g., hoarse voice, hernias, rectal prolapse, joint and skin laxity), and mild mental deficiency with uneven cognitive disabilities
Translocations		
Robertsonian translocation	t(14q21q) translocation	Translocation trisomy 21 (live birth), translocation trisomy 14 (early miscarriage), monosomy 14 or 21 (early miscarriage), a normal chromosome complement (live birth), or a t(14q21q) carrier (live birth).
Reciprocal translocation	t(11;22)(q23.3;q11.2) translocation	Partial trisomy and partial monosomy
Acute promyelocytic leukemia	t(15;17)(q22;q21) reciprocal translocation *PMLI/RARα* oncogene	Pancytopenia (i.e., anemia, neutropenia, and thrombocytopenia), including weakness and easy fatigue, infections of variable severity, and/or hemorrhagic findings (e.g., gingival bleeding, ecchymoses, epistaxis, or menorrhagia), and bleeding secondary to disseminated intravascular coagulation. A rapid cytogenetic diagnosis of this leukemia is essential for patient management because these patients are at an extremely high risk for stroke
Chronic myeloid leukemia	t(9;22)(q34;q11.2) reciprocal translocation Philadelphia chromosome *ABL/BCR* oncogene	Systemic symptoms (e.g., fatigue, malaise, weight loss, excessive sweating), abdominal fullness, bleeding episodes due to platelet dysfunction, abdominal pain may include left upper quadrant pain, early satiety due to the enlarged spleen, tenderness over the lower sternum due to an expanding bone marrow, and the uncontrolled production of maturing granulocytes, predominantly neutrophils, but also eosinophils and basophils

(continued)

t a b l e **11-1** *(continued)*

Cytogenetic Disorder	Chromosomal Defect	Clinical Features
Isochromosomes		
Isochromosome Xq	46, +i(Xq) Centromere divides transversely	Found in 20% of females with **Turner syndrome**, usually as a mosaic cell line along with a 45,X cell line (i.e.,45,X/46, +i[Xq])
Isochromosome 12p	47, +i(12p) Centromere divides transversely	Testicular germ cell tumors Pallister-Killian syndrome: mental retardation, loss of muscle tone, streaks of skin with hypopigmentation, high forehead, coarse facial features, wide space between the eyes, broad nasal bridge, highly arched palate, fold of skin over the inner corner of the eyes, large ears, joint contractures, and cognitive delays
Chromosome Breakage		
Xeroderma pigmentosa	Nucleotide excision repair enzymes; 9q22.3, 3p25 *XPA, XPC* genes	Sunlight (UV radiation) hypersensitivity with sunburnlike reaction, severe skin lesions around the eyes and eyelids, and malignant skin cancers (basal and squamous cell carcinomas and melanomas) whereby most individuals die by 30 years of age
Ataxia-telangiectasia	DNA recombination repair enzymes; 11q22 *ATM* gene	Ionizing radiation hypersensitivity, cerebellar ataxia with depletion of Purkinje cells, progressive nystagmus, slurred speech, oculocutaneous telangiectasia initially in the bulbar conjunctiva followed by ear, eyelid, cheeks, and neck, immunodeficiency, and death in the second decade of life. A high frequency of structural rearrangements of chromosomes 7 and 14 is the cytogenetic observation with this disease
Fanconi anemia	DNA recombination repair enzymes; 16q24 *FA-A* gene	DNA crosslinking agent hypersensitivity, short stature, hypopigmented spots, café-au-lait spots, hypogo- nadism, microcephaly, hypoplastic or aplastic thumbs, renal malformation including unilateral aplasia or horseshoe kidney, acute leukemia, progressive aplastic anemia, head and neck tumors, medulloblas- toma, and is the most common form of congenital aplastic anemia
Bloom syndrome	DNA repair enzymes 15q26 *BLM* gene	Hypersensitivity to DNA-damaging agents, long, narrow face, erythema with telangiectasias in butterfly distribution over the nose and cheeks, high-pitched voice, small stature, small mandible, protuberant ears, absence of upper lateral incisors, well-demarcated patches of hypopigmentation and hyperpigmentation, immunodeficiency with decreased IgA, IgM, and IgG levels, and predisposition to several types of cancers
Hereditary nonpolyposis colorectal cancer	DNA mismatch repair enzymes 3p21.3,2p22, 2p16,7p22 *MLH1, MSH2,MSH6, PMS2* genes	Onset of colorectal cancer at a young age, high frequency of carcinomas proximal to the splenic flexure, multiple synchronous or metachronous colorectal cancers, and presence of extracolonic cancers (e.g., endometrial and ovarian cancer, adenocarcinomas of the stomach, small intestine, and hepatobiliary tract), and, accounts for 3%–5% of all colorectal cancers

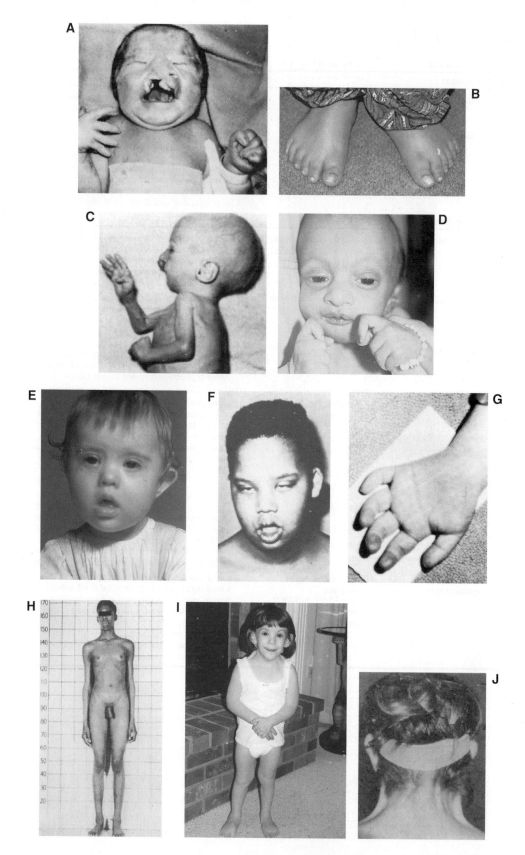

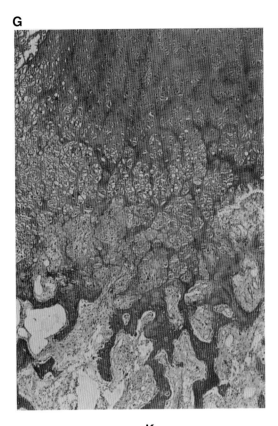

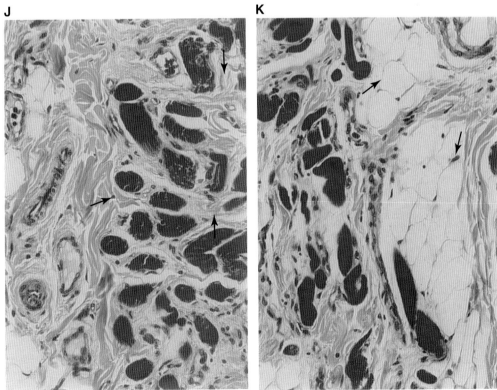

FIGURE 4-2. Selected photographs of Mendelian inherited disorders. **(G)** Light micrograph shows a wide epiphyseal growth plate where the chondrocytes in the zone of proliferation do not form neatly arranged stacks but instead are disorganized into irregular nests. **(J)** Light micrograph shows fibrosis of the endomysium (arrows) surrounding the individual skeletal muscle cells. **(K)** Light micrograph shows the replacement of skeletal muscle cells by adipocytes (arrows) in the later stages of the disorder, which causes pseudohypertrophy.

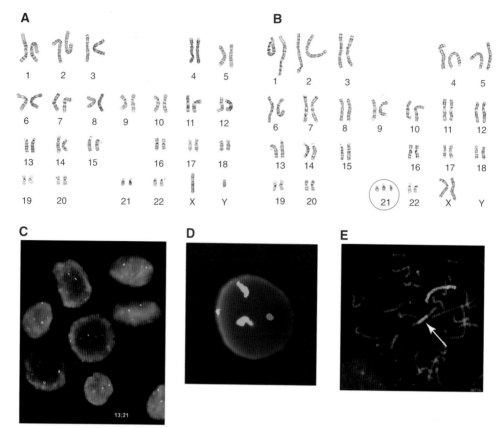

FIGURE 10-1. Karyotypes and chromosomal morphology. (A) G-banding of metaphase chromosomes with only minimal separation of the sister chromatids are shown arranged in a karyotype. Chromosomes 1 through 3 consist of the largest metacentric chromosomes. Chromosomes 4 and 5 are slightly smaller and submetacentric. Chromosomes 6 through12 are arranged in order of decreasing size with the centromere moving from a metacentric position to a submetacentric position. Chromosomes 13 through 15 are medium sized and acrocentric. Chromosomes 16 through18 are smaller and metacentric. Chromosomes 19 and 20 are even smaller and metacentric. Chromosomes 21 and 22 are the smallest chromosomes and acrocentric. The X chromosome is similar to chromosomes 6 through12. The Y chromosome is similar to chromosomes 21 and 22. **(B) Karyotype of Down syndrome.** G-banding of metaphase chromosomes with only minimal separation of the sister chromatids are shown arranged in a karyotype. Note the three chromosomes 21 (circle). **(C) FISH for Down syndrome.** FISH using a probe for chromosome 21 (red dots) shows that each cell contains three red dots indicating trisomy 21. The green dots represent a control probe for chromosome 13. **(D) FISH for sex determination.** FISH using a probes for the X chromosome (green) and the Y chromosome (red) shows that a cell that contain one green dot and one red dot indicating the male sex. The two blue areas represent a control probe for chromosome 18. **(E) Chromosome painting.** Chromosome painting using paints for chromosome 4 (green) and chromosome 14 (red) shows a chromosomal rearrangement between chromosomes 4 and 14 (chromosome with green and red staining; arrow). *(continued)*

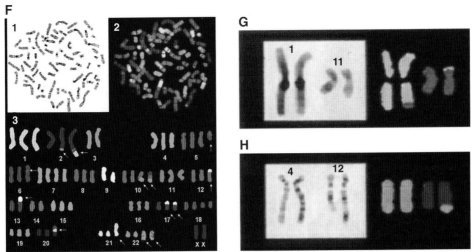

FIGURE 10-1. *(Continued)* **(F) Spectral karyotyping of a chronic myelogenous leukemia cell line demonstrating a complex karyotype with several structural and numerical chromosome aberrations. (F1)** A metaphase cell showing the G-banding pattern. **(F2)** The same metaphase cell as in F1 showing the spectral display pattern. **(F3)** The same metaphase cell as in F1 and F2 arranged as a karyotype and stained with the spectral karyotyping colors. Arrows indicate structural chromosome aberrations involving two or more different chromosomes. **(G) Spectral karyotyping.** Spectral karyotyping using paints for chromosome 1 (yellow) and chromosome 11 (blue) shows a balanced reciprocal translocation between chromosomes 1 and 11, t(1q11p). A balance translocation means that there is no loss of any chromosomal segment during the translocation. This forms two derivative chromosomes each containing a segment of the other chromosome from the reciprocal exchange. **(H) Spectral karyotyping.** Spectral karyotyping using paints for chromosome 4 (blue) and chromosome 12 (red) shows an unbalanced reciprocal translocation between chromosomes 4 and 12, t(4q12q). An unbalanced translocation means that there is loss of a chromosomal segment during the translocation. In this case, the chromosomal segment 12 is lost.

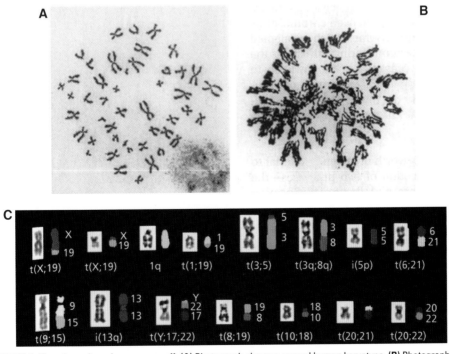

FIGURE 16-1. Karyotype chaos in a cancer cell. (A) Photograph shows a normal human karyotype. **(B)** Photograph shows an abnormal human karyotype due to a mutation involving the RAD 17 checkpoint protein, which plays a role in the cell cycle. This mutation results in a re-replication of already replicated DNA and an abnormal karyotype. **(C)** Spectral karyotyping (24-color chromosome painting) shows twelve chromosome translocations (t) and two isochromosomes in a human urinary bladder carcinoma. See Color Plate.

A2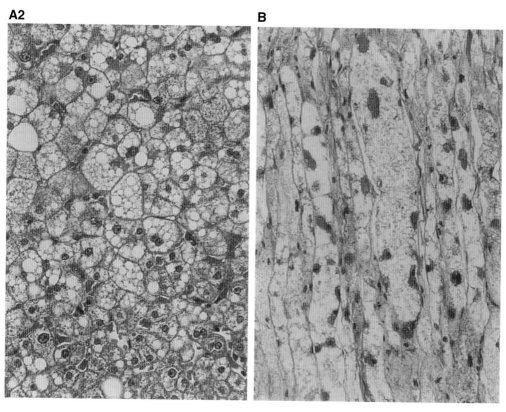

B

F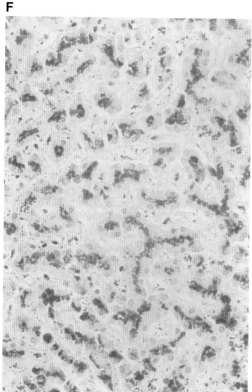

FIGURE 12-1. (A2) Glycogen storage disease type I (von Gierke). (A2) Light micrograph of a liver biopsy shows hepato-cytes with a pale, clear cytoplasm due to the large amounts of accumulated glycogen that is extracted during histological processing. **(B) Glycogen storage disease type V (McArdle).** Light micrograph of a skeletal muscle biopsy shows muscle cells with a pale, clear cytoplasm due to the large amounts of accumulated glycogen that is extracted during histological processing. **(F) Hemochromatosis.** Light micrograph of a liver biopsy stained with Prussian blue shows hepatocytes with a heavily stained cytoplasm to the large amounts of accumulated iron.

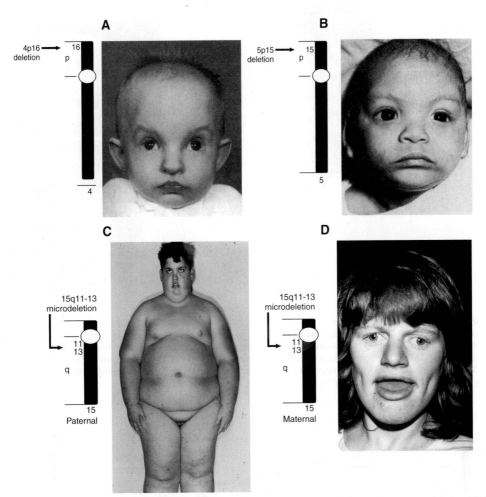

FIGURE 11-4. Structural chromosomal abnormalities (deletion/microdeletions) (A) Chromosome 4p deletion (Wolf-Hirschhorn syndrome). The deletion at 4p16 is shown on chromosome 4. A photograph of a 5-year-old boy with Wolf-Hirschhorn syndrome showing a prominent forehead and broad nasal root ("Greek warrior helmet"), short philtrum, down-turned mouth, and severe mental retardation (IQ = 20). **(B)** Chromosome 5p deletion (Cri du chat; cat cry syndrome). The deletion at 5p15 is shown on chromosome 5. A photograph of an infant with Cri du chat showing round facies, microcephaly, and mental retardation. **(C) Prader-Willi syndrome.** The microdeletion at 15q11-13 is shown on chromosome 15 inherited from the father (paternal). A photograph of a 10-year-old boy with Prader-Willi syndrome showing hypogonadism, hypotonia, obesity, short stature, and small hands and feet. **(D) Angelman syndrome (happy puppet syndrome).** The microdeletion at 15q11-13 is shown on chromosome 15 inherited from the mother (maternal). A photograph of a young woman with Angelman syndrome showing a happy disposition with inappropriate laughter and severe mental retardation (only 5 to 10 word vocabulary). *(continued)*

FIGURE 11-3. Numerical chromosomal abnormalities (aneuploidy). (A,B) Trisomy 13 (Patau syndrome). The key features of Trisomy 13 are microcephaly with sloping forehead, scalp defects, microphthalmia, cleft lip and palate, polydactyly, fingers flexed and overlapping, and cardiac malformations. **(C,D)** Trisomy 18 (Edwards syndrome). The key features of Trisomy 18 are low birth weight, lack of subcutaneous fat, prominent occiput, narrow forehead, small palpebral fissures, low-set and malformed ears, micrognathia, short sternum, and cardiac malformations. **(E,F,G) Trisomy 21 (Down syndrome). (E,F)** Photographs of a young child and boy with Down syndrome. Note the flat nasal bridge, prominent epicanthic folds, oblique palpebral fissures, low-set and shell-like ears, and protruding tongue. Other associated features include: generalized hypotonia, transverse palmar creases (simian lines), shortening and incurving of the fifth fingers (clinodactyly), Brushfield spots, and mental retardation. **(G)** Photograph of hand in Down syndrome showing the simian crease. **(H) Klinefelter syndrome (47,XXY).** Photograph of a young man with Klinefelter syndrome. Note the hypogonadism, eunuchoid habitus, and gynecomastia. **(I,J) Turner syndrome (45,X).** Photograph of a 3-year-old girl with Turner syndrome. Note the webbed neck due to delayed maturation of lymphatics, short stature, and broad shield chest.

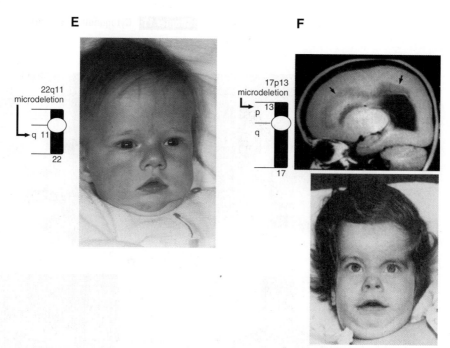

FIGURE 11-4. *(continued)* **(E) DiGeorge syndrome.** The microdeletion at 22q11 is shown on chromosome 22. A photomicrograph of a young infant with craniofacial defects (e.g., hypertelorism, microstomia) along with partial or complete absence of the thymus gland. **(F) Miller-Dieker syndrome (agyria, lissencephaly).** The microdeletion at 17p13.3 is shown on chromosome 17. MRI (top figure) shows a complete absence of gyri in the cerebral hemispheres. The lateral ventricles are indicated by the arrows. A photograph of a young girl with Miller-Dieker syndrome showing small, anteverted nose, long philtrum, and thin prominent upper lip.

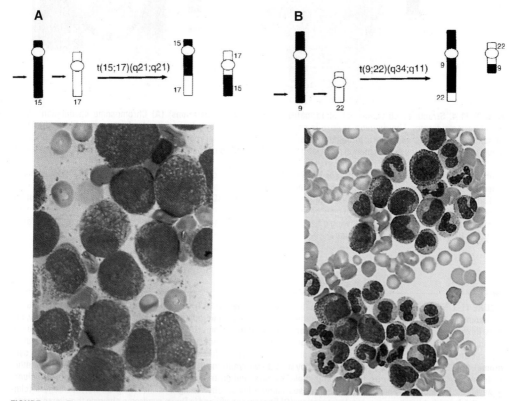

FIGURE 11-5. Translocations. (A) Acute promyelocytic leukemia t(15;17)(q21;q21). The translocation between chromosomes 15 and 17 is shown. A photomicrograph of acute promyelocytic leukemia showing abnormal promyelocytes with their characteristic pattern of heavy granulation and bundle of Auer rods. **(B) Chronic myeloid leukemia t(9;22)(q34;q11).** The translocation between chromosomes 9 and 22 is shown. A photomicrograph of chronic myeloid leukemia showing marker granulocytic hyperplasia with neutrophilic precursors at all stages of maturation. Erythroid (red blood cell) precursors are significantly decreased with none shown in this field.

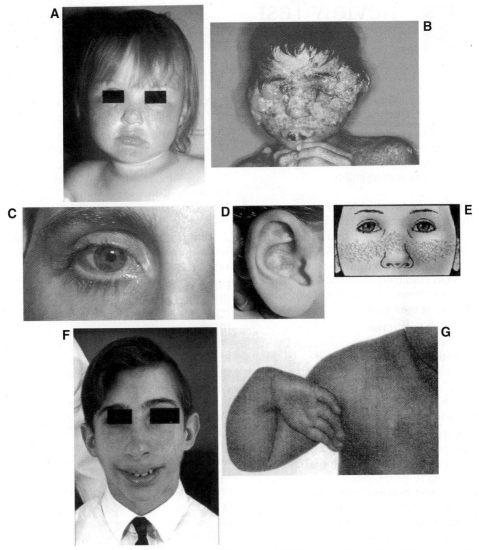

FIGURE 11-6. Chromosome breakage. (A,B) Xeroderma pigmentosa **(C,D,E)** Ataxia-telangiectasia **(F)** Fanconi syndrome **(G)** Bloom syndrome.

Review Test

1. Which of the following patients should be offered cytogenetic studies?

(A) parents of a child with trisomy 21
(B) parents of a child with Turner syndrome
(C) a 37-year-old woman who is pregnant
(D) parents of a normal child

2. Amniocentesis is performed on a patient at 16 weeks' gestation because of her age (she is 36). The final report to the physician says that the fetus has a 45X/46,XX karyotype, with the 45,X cell line making up 90% of the cells examined. The fetus will most likely have phenotypic features of which of the following syndromes?

(A) Fragile X syndrome
(B) Turner syndrome
(C) Down syndrome
(D) Angelman syndrome

3. Which of the following is one of the most common causes of Prader-Willi syndrome?

(A) a microdeletion on the maternal chromosome 15
(B) a microdeletion on the paternal chromosome 15
(C) a microdeletion on the maternal chromosome 22
(D) a microdeletion on the paternal chromosome 22

4. Which one of the following Robertsonian translocation carriers has the greatest risk of having an abnormal child?

(A) 45,XX,t(14;15)
(B) 45,XY,t(15;22)
(C) 45,XX,t(13;21)
(D) 45,XY,t(14;22)

5. Which of the following is the main risk to children of inversion carriers?

(A) Down syndrome
(B) duplications or deletions
(C) chronic myelogenous leukemia
(D) Robertsonian translocations

6. Which one of the following is an indication that you should offer a patient cytogenetic studies?

(A) family history of Huntington disease
(B) family history of unexplained miscarriages and mental retardation
(C) family history of tall stature
(D) family history of cystic fibrosis

7. A tall male with gynecomastia and small testes should have a cytogenetic study to rule out which of the following?

(A) XYY syndrome
(B) Klinefelter syndrome
(C) Fragile X syndrome
(D) Turner syndrome

8. A woman comes to clinic because of her family history of tetralogy of Fallot (a conotruncal heart defect). Her father was born with a heart defect, has immunity problems, and schizophrenia. Her brother has cleft palate and a heart defect as well. The patient was studied cytogenetically and found to have a microdeletion of 22q11.2 by FISH. What is the best estimate of her recurrence risk for a future pregnancy?

(A) 2%–3%
(B) 5%–6%
(C) 10%
(D) 50%
(E) 100%

9. You see a 4-year-old boy in clinic whom you believe has Prader-Willi syndrome. You request cytogenetic studies and the child is found to have an unbalanced 14;15 translocation. Fluorescent *in situ* hybridization (FISH) confirms that the Prader-Willi/Angelman area on chromosome 15 is deleted. You request cytogenetic studies of the parents and one of them is found to have a balanced translocation. Which of the following are the most likely cytogenetic findings?

(A) The father has a balanced 14;15 translocation.

(B) The father's Prader-Willi/Angelman locus is found by FISH to be deleted.
(C) The mother has a balanced 14;15 translocation.
(D) The mother's Prader-Willi/Angelman locus is found by FISH to be deleted.

10. Which of the following is the most common cause of Down syndrome?

(A) Robertsonian translocations
(B) 21;21 balanced reciprocal translocation
(C) nondisjunction in mitosis
(D) nondisjunction in meiosis

11. A chromosome deletion results in which of the following?

(A) a chromosome monosomic for the deleted area
(B) a chromosome disomic for the deleted area
(C) a chromosome trisomic for the deleted area
(D) a chromosome tetrasomic for the deleted area

12. In the 9;22 translocation characteristic of chronic myeloid leukemia (CML), the Philadelphia chromosome has which of the following characteristics?

(A) It is deleted for the abl proto-oncogene.
(B) It is deleted for the bcr proto-oncogene.
(C) It has the abl/bcr fusion gene generated by the 9;22 translocation.
(D) It is deleted for the abl/bcr fusion gene.

13. Jane and her husband Charlie have a phenotypically normal female child with a balanced Robertsonian translocation between chromosomes 13 and 21. How many chromosomes does the child have?

(A) 46
(B) 47
(C) 45
(D) 48

14. Jane and Charlie from question 13 wish to have more children. What should their physician recommend as their next course of action?

(A) Recommend that they have no more children because of the risk of having an abnormal child.
(B) Recommend that Jane be studied to determine if she is a carrier of the Robertsonian translocation.
(C) Recommend that Charlie be studied to determine if he is a carrier of the Robertsonian translocation.
(D) Recommend that both Jane and Charlie be studied to determine if one of them is a carrier of the Robertsonian translocation.

15. The greatest risk of having a child with a chromosome abnormality will occur with which one of the following?

(A) a couple who had a child with a *de novo* (spontaneously occurring) unbalanced translocation between chromosomes 2 and 5
(B) a couple who had a child with an unbalanced translocation between chromosomes 2 and 5, and the father was identified as a carrier of a balanced 2;5 translocation
(C) a couple who had a child with Down syndrome
(D) a couple who have had no children

16. Which of the following karyotypes is most likely to result in a viable (capable of being born alive) outcome?

(A) 47,XYY
(B) 47,+16
(C) 69,XXX
(D) 47,XY,+18

17. Which of the following is the best estimate of the chance that a child produced by the union of a female carrier of a 21;21 Robertsonian translocation carrier and a karyotypically normal male will have Down syndrome?

(A) 0%
(B) 5%
(C) 15%
(D) 100%

18. A nondisjunction of chromosome 21 in meiosis II in a male would yield which combination of the following gametes?

(A) one sperm with two chromosome 21s, the rest with one
(B) one sperm with no 21s, three with two 21s
(C) one sperm with two chromosome 21s, one with no chromosome 21s and two sperm with one chromosome 21
(D) all of the sperm would have one chromosome 21

Answers and Explanations

1. **The answer is (C).** A 37-year-old woman has about a 1% risk to have a child with a chromosome abnormality and should be offered amniocentesis to detect chromosome abnormalities in the fetus. Trisomy 21 and Turner syndrome occur spontaneously so cytogenetic studies of the parents would not provide any information on future risk. Parents of a normal child have the population risk for having a child with a chromosome abnormality so there is no indication for offering the test.

2. **The answer is (B).** Many patients with Turner syndrome are mosaics, that is, they have two or more cell lines with different karyotypes. Although there is a normal, 46,XX cell line present, the majority of cells have the 45,X Turner syndrome karyotype and thus some phenotypic features of Turner syndrome can be expected.

3. **The answer is (B).** The most common cause of Prader-Willi syndrome is a microdeletion in the area of the long arm of chromosome 15 between bands q11 and q13. This area of chromosome 15 is genomically imprinted, so the parent of origin for the chromosome determines what syndrome will occur as a result of the deletion. If the deletion is on the chromosome 15 that came from the father, then Prader-Willi syndrome will result. Angelman syndrome occurs if the microdeletion is on the maternal chromosome 15.

4. **The answer is (C).** Carriers of a 13;21 Robertsonian translocation are at risk for having a child with Robertsonian Down syndrome or Robertsonian trisomy 13. All the other Robertsonian translocation carriers have Robertsonian translocations that are lethal when trisomy occurs and most of these conceptions are not even recognized pregnancies. There may be an increased risk of infertility connected with these Robertsonian translocations, but no increased risk of having abnormal children.

5. **The answer is (B).** During meiosis, the inverted chromosome must pair with its homolog in a way that forms a loop. Crossing-over within the inversion loop can result in duplications or deletions of parts of the chromosomes. These duplicated and deleted chromosomes are thus in the gamete resulting from the meiosis and when this unbalanced gamete and a normal gamete fuse, the conceptus will have an unbalanced chromosome complement.

6. **The answer is (B).** A family history of unexplained miscarriages and mental retardation may indicate that a structural chromosome rearrangement is segregating in the family and the miscarriages and mental retardation are the result of inheriting unbalanced segregants. Cytogenetic testing is not indicated for the other choices.

7. **The answer is (B).** Klinefelter syndrome, which is the result of a 47,XXY chromosome constitution, is characterized, among other things, by tall stature, gynecomastia, and small testes. This combination of features is not seen in the other choices.

8. **The answer is (D).** Because there is one normal chromosome 22 and one deleted chromosome 22, there is a 50% chance of passing one or the other on with each pregnancy.

9. **The answer is (A).** FISH analysis of the child's chromosomes showed that the Prader-Willi/Angelman (PWA) locus on chromosome 15 was not present. Any unbalanced rearrangement of chromosome 15 that would have been inherited to cause Prader-Willi in the child would have to come from the father, since it is the paternally inherited deletion of chromosome 15 that causes most cases of Prader-Willi syndrome. The father could not be

carrying a deletion of the PWA locus or he would have Prader-Willi syndrome, which would certainly be identified by you, the physician. Individuals with Prader-Willi and Angelman syndromes do not reproduce.

10. **The answer is (D).** Although certain Robertsonian translocations involving chromosome 21 and other translocations involving chromosome 21 can result in Down syndrome, the most common cytogenetic finding in Down syndrome is three copies of chromosome 21, or trisomy for chromosome 21. Trisomy 21 is caused by nondisjunction of chromosome 21 during meiosis. A nondisjunction of chromosome 21 in mitosis is rare and would lead to mosaicism for trisomy 21.

11. **The answer is (A).** In a deletion, a portion of the chromosome is lost, leaving only one copy of that area on the homologous chromosome. Since there is only one copy left on the normal homolog, that chromosome is monosomic for the deleted area.

12. **The answer is (C).** The abl proto-oncogene on chromosome 9 and the bcr proto-oncogene on chromosome 22 are fused by the 9;22 translocation. Deletions of these genes are not associated with CML.

13. **The answer is (C).** Because a Robertsonian translocation leads to the fusion of two chromosomes, in this case chromosomes 13 and 21, there is one less chromosome in the karyotype as a result. The chromosome number thus goes from 46 to 45.

14. **The answer is (D).** Both parents should be studied because either parent could be a carrier. If neither parent is a carrier, this would mean that there was little risk of having another child with a chromosome abnormality. Carriers of Robertsonian translocations can have normal children, children who are balanced carriers like themselves or children with chromosome abnormalities. In the case of a 13;21 Robertsonian translocation, the risk of having an abnormal child would be ~5% if the father is a carrier, and 15% if the mother is a carrier.

15. **The answer is (B).** Carriers of some balanced translocations have a significant risk of having a child with a chromosome abnormality. Because the couple has already had an abnormal child, this indicates that viable, abnormal outcomes are possible with this particular balanced translocation and the couple has a significantly elevated risk of having it happen again.

16. **The answer is (A).** A 47,XYY karyotype is usually only detected incidentally to some other indication for study since there is not an abnormal phenotype associated with it. Males with this karyotype are just as likely to be viable as those with a normal 46,XY karyotype. A 47,XX or XY, +18 karyotype can result in a liveborn, but the majority of fetuses with this karyotype spontaneously abort. Triploids (69 chromosomes) are rarely liveborn and even then do not usually survive beyond a couple of hours. A large percentage of first trimester spontaneous abortions have a 47,XX or XY, +16 karyotype.

17. **The answer is (D).** In meiosis, a 21;21 Robertsonian translocation chromosome will go into one of the daughter cells during meiosis I and the other daughter cell will not receive anything. Thus, a 21;21 Robertsonian translation carrier can only produce gametes that are disomic for chromosome 21 or nullisomic for chromosome 21. When an ovum from a carrier female is fertilized with a normal sperm, the union of the sperm with its one copy of chromosome 21 and the ovum with its two copies of chromosome 21 contained in the 21;21 translocation chromosome will result in three copies of chromosome 21 being present in the conceptus and Down syndrome will be the result. The fertilization of a nullisomic ovum, with no copy of chromosome 21, by a normal sperm with its one copy of chromosome 21, will result in a conceptus that is monosomic for chromosome 21 and this is lethal. Thus, there is almost a 100% chance that any child resulting from this union will have Down syndrome.

18. The answer is (B). Because there are no outstanding phenotypic characteristics associated with XYY, people with this karyotype are usually only diagnosed accidentally. Individuals with Fragile X syndrome have, among other phenotypic abnormalities, mental retardation. Turner syndrome is found only in females.

19. The answer is (C). In nondisjunction at meiosis II one of the two daughter cells resulting from cell division in meiosis I would proceed to divide normally during meiosis II, resulting in two normal daughter cells. The nondisjunction of the paired chromosome 21s in the other meiosis I daughter cell would during meiosis II would lead to both chromosome 21's going to one daughter cell and no chromosome 21s going to the other daughter cell.

I. INTRODUCTION

A. Metabolic reactions within various biochemical pathways are controlled by **enzymes** that increase the reaction rate by a million fold. In general, metabolic genetic disorders are caused by mutations in genes that encode for enzymes of various biochemical pathways.

B. Most metabolic genetic disorders are **autosomal recessive disorders** (see Chapter 4-II) whereby individuals with two mutant alleles (homozygous recessive) demonstrate clinically apparent, phenotypic errors in metabolism. A heterozygote is generally normal because the one normal allele produces enough enzymatic activity to maintain normal metabolism.

C. The parents of a proband are obligate heterozygotes whereby each parent carries one mutant allele and is asymptomatic.

II. METABOLIC GENETIC DISORDERS INVOLVING CARBOHYDRATE PATHWAYS

A. Galactosemia (GAL).
1. GAL is an autosomal recessive genetic disorder caused by various **missense mutations** in the **GALT** gene on **chromosome 9p13** for **galactose-1-phosphate uridylyltransferase (GALT)** which catalyzes the reaction galactose-1-phosphate → glucose-1-phosphate.
2. The various missense mutations result either in a **normal glutamine → arginine** substitution at position 188 (Q188R) prevalent in northern Europe; a **normal serine → leucine** substitution at position 135 (S135L) prevalent in Africa; or a **normal lysine → asparagine** substitution at position 285 (K285N) prevalent in Germany, Austria, and Croatia.
3. The **Duarte variant allele** is caused by a missense mutation which results in a normal asparagine → aspartate substitution at position 314 (N314D) which imparts instability to GALT whereby affected individuals have 5% to 20% GALT activity compared to normal individuals.
4. **Prevalence.** The prevalence of GAL is 1/30,000 births.
5. **Clinical features include:** feeding problems in the newborn; failure to thrive, hypoglycemia, hepatocellular damage, bleeding diathesis, jaundice, and hyperammonemia; sepsis with *E. coli*, shock, and death may occur if the galactosemia is not treated; galactosemia is one of the conditions tested for on newborn screens in most states.

B. Asymptomatic Fructosuria (AF; or Essential Fructosuria).
1. AF is an autosomal recessive genetic disorder caused by a mutation in the **KHK** gene on **chromosome 2p23.3-p23.2** for **ketohexokinase (or fructokinase)** which catalyzes the reaction fructose → fructose-1-phosphate.
2. **Clinical features include:** asymptomatic presence of fructose in the urine.

C. Hereditary Fructose Intolerance (HFI; Fructosemia).
1. HFI is an autosomal recessive genetic disorder caused by a mutation in the ***ALDOB* gene** on **chromosome 9q21.3-q22.2** for **fructose 1-phosphate aldolase B**, which catalyzes the reaction fructose 1-phosphate → dihydroxyacetone phosphate + D-glyceraldehyde.
2. The most likely mechanism causing the clinical features of HFI is that the PO_4^{3-} group gets sequestered on fructose and therefore is not available for ATP synthesis.
3. **Prevalence.** The prevalence of HFI is 1/20,000 births.
4. **Clinical features include:** failure to thrive, fructosuria, hepatomegaly, jaundice, aminoaciduria, metabolic acidosis, lactic acidosis, low urine ketones, recurrent hypoglycemia and vomiting at the age of weaning when fructose or sucrose (a disaccharide that is hydrolyzed to glucose and fructose) is added to the diet; infants and adults are asymptomatic until they ingest fructose or sucrose.

D. Lactose Intolerance (LI; Lactase Nonpersistence; Adult-Type Hypolactasia).
1. LI is an autosomal recessive genetic disorder associated with short tandem repeat polymorphisms (STRPs) in the promoter region that affects transcriptional activity of the ***LCT* gene** on **chromosome 2q21** for **lactase-phlorizin hydrolase** which catalyzes the reaction lactose → glucose + galactose.
2. These STRPs in the human population lead to two distinct phenotypes: **lactase persistent** individuals and **lactase nonpersistent** individuals.
3. All healthy newborn children up to the age of ≈5 to 7 years of age have high levels of lactase-phlorizin hydrolase activity so that they can digest large quantities of lactose present in milk.
4. Northern European adults (particularly Scandinavian) retain high levels of lactase-phlorizin activity and are known as **lactase persistent** and therefore **lactose tolerant.**
5. However, a majority of the world's adults (particularly in Africa and Asia) lose the high levels of lactase-phlorizin activity and are known as **lactase nonpersistent** and therefore **lactose intolerant.**
6. **Clinical findings of lactose intolerance include:** diarrhea, crampy abdominal pain localized to the periumbilical area or lower quadrant, flatulence, nausea, vomiting, audible borborygmi, stools that are bulky, frothy, and watery, and bloating after milk or lactose consumption.

E. Glycogen Storage Disease Type I (GSDI; von Gierke).
1. **GSDIa** is an autosomal recessive genetic disorder caused by ≈85 different mutations in the ***G6PC* gene** on **chromosome 17q21** for **glucose-6-phosphatase**, which catalyzes the reaction glucose-6-phosphate → glucose + phosphate.
2. **GSDIb** is an autosomal recessive genetic disorder caused by ≈78 different mutations in the ***SLC37A4* gene** on **chromosome 11q23** for **glucose-6-phosphate translocase**, which transports glucose-6-phosphate into the lumen of the endoplasmic reticulum.
3. GSDIa is commonly (32% of cases in the Caucasian population and 93% to 100% of cases in the Jewish population) caused by a **missense mutation** which results in a **normal arginine → cysteine** substitution at position 83 (R83C). GSDIb is commonly (15% of cases in the Caucasian population and 30% of cases in the German population) caused by a **missense mutation** which results in a **normal glycine → cysteine** substitution at position 339 (G339C).
4. **Prevalence.** The prevalence of GSDI is 1/100,000 births.
5. **Clinical features include:** accumulation of glycogen and fat in the liver and kidney resulting in hepatomegaly and renomegaly, severe hypoglycemia, lactic acidosis, hyperuricemia, hyperlipidemia, hypoglycemic seizures, doll-like faces with fat cheeks, relatively thin extremities, short stature, protuberant abdomen, and neutropenia with recurrent bacterial infections.

F. Glycogen Storage Disease Type V (GSDV; McArdle Disease).
1. GSDV is an autosomal recessive genetic disorder caused by ≈46 different mutations in the ***PYGM* gene** on **chromosome 11q13** for **muscle glycogen phosphorylase**, which initiates glycogen

breakdown by removing α1,4glucosyl residues from the outer branches of glycogen with liberation of glucose-1-phosphate.

2. GSDV is commonly caused by either a nonsense mutation which results in a **normal arginine → nonsense** at position 49 (R49X) causing a premature STOP codon (90% of cases in European and US populations) or a missense mutation which results in a **normal glycine → serine** substitution at position 204 (G204S; 10% of cases in European and US populations).

3. **Prevalence.** The prevalence of GSDV is 1/100,000 births.

4. **Clinical features include:** exercise-induced muscle cramps and pain, "second wind" phenomenon with relief of myalgia and fatigue after a few minutes of rest, episodes of myoglobinuria, increased resting basal serum creatine kinase (CK) activity, onset typically occurs around 20 to 30 years of age; clumsiness, lethargy, slow movement, and laziness in preadolescents.

G. **Other Glycogen Storage Diseases.** These include: glycogen storage disease type II (GSDII; Pompe); glycogen storage disease type IIIa (GSDIIIa; Cori); glycogen storage disease type IV (GSDIV; Andersen); glycogen storage disease type VI (GSDVI; Hers); and glycogen storage disease type VII (GSDVII; Tarui).

III. METABOLIC GENETIC DISORDERS INVOLVING AMINO ACID PATHWAYS

A. **Phenylalanine Hydroxylase (PAH) Deficiency (or PKU).**
1. PAH deficiency is an autosomal recessive genetic disorder caused by a mutation in the *PAH* gene on **chromosome 12q23.2 for phenylalanine hydroxylase** which catalyzes the reaction phenylalanine → tyrosine.

2. PAH deficiency is caused by missense (most common; 62% of cases); small deletion (13% of cases); RNA splicing (11% of cases); silent (6% cases); nonsense (5% of cases); or insertion (2% of cases) mutations.

3. PAH deficiency results in an intolerance to the dietary intake of phenylalanine (an essential amino acid). This produces a variability in metabolic phenotypes including **classic phenylketonuria (PKU), non-PKU hyperphenylalaninemia,** and **variant PKU.** This variability in metabolic phenotypes is caused primarily by different mutations in the PAH gene that result in variations in the kinetics of phenylalanine uptake, permeability of the blood–brain barrier, and protein folding.

4. Classic PKU is associated with the complete absence of PAH and is the most severe of the three types of PAH deficiency.

5. **Prevalence.** The prevalence of PAH deficiency is 1/10,000 births in the Caucasian population and 1/200,000 in the Ashkenazi Jewish population. Since the advent of universal newborn screening, symptomatic classic PKU is rarely seen.

6. **Clinical features of classic PKU include:** no physical signs are apparent in neonates with PAH deficiency; diagnosis is based on detection of elevated plasma phenylalanine concentration (>1,000 umol/L for classic PKU) and normal BH_4 cofactor metabolism; a dietary phenylalanine tolerance of <500 mg/day; untreated children with classic PKU show impaired brain development, microcephaly, epilepsy, severe mental retardation, behavioral problems, depression, anxiety, musty body odor, and skin conditions like eczema.

B. **Hereditary Tyrosinemia Type I (TYRI).**
1. **TYRI** is an autosomal recessive genetic disorder caused by a mutation in the *FAH* gene on **chromosome 15q23-q25 for fumarylacetoacetate hydrolase,** which catalyzes the reaction fumarylacetoacetic acid → fumarate + acetoacetate.

2. **TYRI** is caused by either a missense mutation, which results in a **normal proline → leucine** substitution at position 261 (P261L) or RNA splicing mutations. The P261L mutation

accounts for 100% of cases in the Ashkenazi Jewish population. The P261L and some RNA splicing mutations account for 60% of cases in the US population.

3. **Prevalence.** The prevalence of TYRI is 1/120,000 births.
4. **Clinical features include:** diagnosis is based on detection of elevated plasma succinylace-tone concentration; elevated plasma tyrosine, methionine, and phenylalanine concentra-tions; elevated urinary tyrosine metabolite (e.g., hydroxyphenylpyruvate) concentration; elevated urinary δ-aminolevulinic acid; cabbage-like odor; untreated children with HTI show severe liver dysfunction, renal tubular dysfunction, growth failure, and rickets.

C. Maple Syrup Urine Disease (MSUD).

1. MSUD is an autosomal recessive genetic disorder cause by >60 different mutations in either the **BCKDHA gene** on **chromosome 19q13.1-q13.2** for the **E1α subunit of the branched-chain ketoacid dehydrogenase complex (BCKD)**, the **BCKDHB gene** on **chromosome 6q14** for the **E1ß subunit of BCKD**, or the **DBT gene** on **chromosome 1p31** for the **E2 subunit of BCKD** all of which catalyze the second step in the degradation of branched-chain amino acids (e.g., leucine, isoleucine, and valine).
2. The BCKD enzyme is an enzyme complex found in the mitochondria.
3. **Prevalence.** The prevalence of MSUD is 1/185,000 births.
4. **Clinical features include:** untreated children with MSUD show maple syrup odor in ceru-men 12 to 24 hours after birth, elevated plasma branched-chain amino acid concentra-tion, ketonuria, irritability, poor feeding by 2 to 3 days of age, deepening encephalopathy including lethargy, intermittent apnea, opisthotonus, and stereotyped movements like "fencing" and "bicycling" by 4 to 5 days of age; acute leucine intoxication (leucinosis) associated with neurological deterioration due to the ability of leucine to interfere with the transport of other large neutral amino acids across the blood–brain barrier, thereby reducing the amino acid supply to the brain.

IV. METABOLIC GENETIC DISORDERS INVOLVING LIPID PATHWAYS

A. Medium-Chain Acyl-coenzyme A Dehydrogenase (MCAD) Deficiency.

1. MCAD deficiency is an autosomal recessive genetic disorder caused by ≈45 different mutations in the **ACADM gene** on **chromosome 1p31** for **medium-chain acyl-coenzyme A dehydrogenase (MCAD)** which catalyzes the initial dehydrogenation of acyl-CoAs with a fatty acid chain length of 4 to 12 carbon atoms.
2. The *ACADM* gene is a nuclear gene that codes for MCAD enzyme, which is active in the mitochondria and part of the mitochondrial fatty acid ß-oxidation pathway. A defect in MCAD leads to an accumulation of medium-chain fatty acids, which are further metabo-lized to glycine-esters, carnitine-esters, and dicarboxylic acids (all of which are detectable in blood, urine, and bile).
3. The mitochondrial fatty acid β-oxidation pathway normally fuels hepatic ketogenesis, which is a major source of energy when hepatic glycogen stores are depleted during pro-longed fasting or high energy demands.
4. MCAD deficiency is caused by a missense mutation which results in a **normal lysine →** **glutamate** substitution at position 304 (K304E) prevalent in the Northern European popu-lation.
5. **Prevalence.** The prevalence of MCAD is 1/15,700 births in the US population. MCAD is especially prevalent in Caucasians of Northern European descent.
6. **Clinical features include:** hyperketotic hypoglycemia, vomiting, and lethargy triggered by either a common illness (e.g., viral gastrointestinal or upper respiratory tract infections) or prolonged fasting (e.g., weaning the infant from nighttime feedings) which may quickly progress to coma and death; hepatomegaly and acute liver disease; children are normal at birth and present between 3 and 24 months of age; later presentation into adulthood is possible.

B. Smith-Lemli-Opitz (SLO) Syndrome.

1. SLO syndrome is an autosomal recessive genetic disorder caused by >70 different mutations in the **DHCR7 gene** on **chromosome 11q12-q13** for **7-dehydrocholesterol reductase** which catalyzes the last step in cholesterol biosynthesis 7-dehydrocholesterol → cholesterol.

2. SLO syndrome is commonly caused either by a missense mutation which results in a **normal threonine → methionine** substitution at position 93 (T93M), a nonsense mutation which results in a **normal tryptophan → nonsense** at position 151 (W151X) causing a premature STOP codon, or a intron 8 splice acceptor mutation, all of which account for ≈50% of all cases.

3. **Prevalence.** The prevalence of SLO is 1/20,000 to 40,000 births.

4. **Clinical features include:** prenatal and postnatal growth retardation, microcephaly, moderate to severe mental retardation, cleft palate, cardiac defects, underdeveloped external genitalia and hypospadias in males, postaxial polydactyly, Y-shaped 2 to 3 toe syndactyly, downslanting palpebral fissures, epicanthal folds, anteverted nares, and micrognathia.

C. Familial Hypercholesterolemia (FH).

1. FH is an autosomal dominant genetic disorder caused by >400 different mutations in the **LDLR gene** on **chromosome 19p13.1-13.3** for the **low-density lipoprotein receptor** which binds LDL and delivers LDL into the cell cytoplasm.

2. Mutations in the *LDLR* gene are grouped into 6 classes:
   a. **Class 1** mutations prevent LDLR synthesis.
   b. **Class 2** mutations prevent LDLR transport to the cell membrane.
   c. **Class 3** mutations prevent LDL binding to LDLR.
   d. **Class 4** mutations prevent LDL internalization into the cell cytoplasm by coated pits.
   e. **Class 5** mutations prevent LDLR recycling back to the cell membrane after LDL + LDLR dissociation.
   f. **Class 6** mutations prevent LDLR targeting to the apical membrane adjacent to the blood capillaries.

3. Other genes associated with FH include:
   a. FH is also an autosomal dominant genetic disorder caused by a mutation in the **APOB gene** on **chromosome 2p23-p24** for **apolipoprotein B-100** which is a component of LDL and the ligand for LDLR. The prevalence of *APOB* gene homozygotes is 1/1,000,000 births. The prevalence of *APOB* gene heterozygotes is 1/1,000 births in Caucasians of European descent.
   b. FH is also an autosomal dominant genetic disorder caused by a missense, gain-of-function mutations in the **PCSK9 gene** on **chromosome 1p32-p34.1** for **proprotein convertase subtilisin/kexin type 9.** The increased PCSK9 protease activity degrades LDLR leading to hypercholesterolemia. This type of FH is very rare.
   c. The **Tyr142Stop** and **Cys679Stop** nonsense mutations in the **PCSK9 gene** are loss-of-function mutations. The decreased PCSK9 protease activity has been associated with a 40% reduction in LDL cholesterol (i.e., **hypocholesterolemia**) and a 90% reduced risk of coronary artery disease in 2.6% of the African American population.
   d. The **Arg46Leu** mutation in the **PCSK9 gene** is a loss-of-function mutation. The decreased PCSK9 protease activity has been associated with a 15% reduction in LDL cholesterol (i.e., **hypocholesterolemia**) and 50% reduced risk of coronary artery disease in 3.2% of whites in the US population.

4. **Prevalence.** The prevalence of *LDLR* gene homozygotes is 1/1,000,000 births. The prevalence of *LDLR* gene heterozygotes is 1/500 births. Most cases of hypercholesterolemia and hyperlipoproteinemia in the general population are of multifactorial origin.

5. **Clinical features include:** premature heart disease as a result of atheromas (deposits of LDL-derived cholesterol in the coronary arteries); xanthomas (cholesterol deposits in the skin and tendons); arcus lipoides (deposits of cholesterol around the cornea of the eye); homozygote and heterozygote phenotypes are known; homozygotes develop severe symptoms early in life and rarely live past 30 years of age; heterozygotes have plasma cholesterol level twice that of normal.

V. METABOLIC GENETIC DISORDERS INVOLVING THE UREA CYCLE PATHWAY

A. The urea cycle produces the amino acid arginine (this is the only source of endogenous arginine) and clears waste nitrogen resulting from the metabolism of proteins and dietary intake (this is the only pathway for waste nitrogen clearance). The waste nitrogen is converted to ammonia (NH_4) and transported to the liver.

B. The severity of these disorders is influenced by the position of the defective enzyme in the urea cycle pathway and the severity of the enzyme defect (partial activity vs. absent activity).

C. Because the urea cycle is the only pathway for waste nitrogen clearance, clinical symptoms develop very rapidly.

D. Prevalence. The prevalence of urea cycle disorders is 1/30,000 births.

E. Clinical features include: infants initially appear normal but then rapidly develop hyperammonemia, cerebral edema, lethargy, anorexia, hyperventilation or hypoventilation, hypothermia, seizures, neurologic posturing, and coma; in infants with partial enzyme deficiencies, the symptoms may be delayed for months or years, the symptoms are more subtle, the hyperammonemia is less severe, and ammonia accumulation can be triggered by illness or stress throughout life.

F. Metabolic genetic disorders involving the urea cycle pathway include:
  1. Ornithine transcarbamylase (OTC) deficiency.
    a. OTC deficiency is an X-linked recessive genetic disorder caused by a mutation in the **OTC gene** on **chromosome Xp21.1** for **ornithine transcarbamylase**.
    b. OTC deficiency along with CPSI deficiency and NAGS deficiency are the most severe types of urea cycle disorders. Newborns with OTC deficiency rapidly develop hyperammonemia and these children are always at risk for repeated bouts of hyperammonemia.
    c. OTC can be distinguished from carbamoylphosphate synthetase (CPSI) deficiency by elevated levels of **orotic acid** in OTC individuals.
    d. ≈15% of female carriers develop hyperammonemia during their lifetime and many require chronic medical management.
  2. Other urea cycle disorders. These include: carbamoylphosphate synthetase I (CPSI) deficiency; argininosuccinic acid synthetase (ASS) deficiency (or citrullinemia type I); argininosuccinic acid lyase (ASL) deficiency (or argininosuccinic aciduria); arginase (ARG) deficiency (or hyperargininemia); and N-acetyl glutamate synthetase (NAGS) deficiency.

VI. METABOLIC GENETIC DISORDERS INVOLVING TRANSPORT PATHWAYS

A. Menkes Disease (MND).
  1. MND is an X-linked recessive genetic disorder caused by various mutations in the **ATP7A gene** on **chromosome Xq12-q13** for **Copper-Transporting ATPase 1,** which is a P-type ATPase that transports copper across cell membranes thereby controlling copper homeostasis.
  2. MND is commonly caused by small insertion and deletion mutations (35%), nonsense mutations (20%); RNA splicing mutations (15%), and missense mutations (8%).
  3. These mutations result in low serum concentration of copper (0 to 60 μg/dL vs. 70 to 150 μg/dL normal), low serum concentration of ceruloplasmin (30 to 150 mg/dL vs. 200 to 450 mg/dL normal), a decreased intestinal absorption of copper, an accumulation of copper

in some tissues, and a decreased activity of copper-dependent enzymes (e.g., dopamine ß-hydroxylase critical for catecholamine synthesis or lysyl oxidase).

4. **Prevalence.** The prevalence of MND is 1/1,000,000 births.
5. **Clinical features include:** infants initially appear normal up to 2 to 3 months of age but then develop hypotonia; seizures; failure to thrive; loss of developmental milestones; changes in hair (short, coarse, twisted, lightly pigmented, "steel wool" appearance); jowly facial appearance with sagging cheeks; temperature instability; hypoglycemia; urinary bladder diverticulae; and gastric polyps. Without early treatment with parenteral copper, MND progresses to severe neurodegeneration and death by 7 months → 3 years of age.

B. Wilson Disease (WND).
1. WND is an autosomal recessive genetic disorder caused by >260 mutations in the *ATP7B* **gene** on chromosome **13q14.3-q21.1** for **Copper-Transporting ATPase 2,** which is a P type ATPase expressed mainly in the kidney and liver that plays a key role in incorporating copper into ceruloplasmin and in the release of copper into bile.
2. WND is commonly caused by either a missense mutation which results in a **normal histidine → glutamine** substitution at position 1069 (H1069Q), a missense mutation which results in a **normal arginine → leucine** substitution at position 778 (R778L), or a **15 base pair deletion** in the promoter region.
3. The H1069Q mutation accounts for 45% of cases in the European population. The R778L mutation accounts for 60% of cases in the Asian population. The 15 base pair deletion mutation is common in the Sardinian population.
4. These mutations result in **high hepatic concentration of copper** (>250 μg/g dry weight vs. <55 μg/g dry weight normal); high urinary excretion of copper (>0.6 umol/24 hours); and **damage of various tissues due to excessive accumulation of copper.**
5. **Prevalence.** The prevalence of WND is 1/30,000 in most populations. The prevalence is 1/10,000 in Chinese and Japanese populations.
6. **Clinical features include:** symptoms occur in individuals from 3 to 50 years of age; recurrent jaundice; hepatitislike illness; fulminant hepatic failure; tremors; poor coordination; loss of fine motor control; chorea; masklike facies; rigidity; gait disturbance; depression; neurotic behaviors; **Kayser-Fleischer rings** (deposition of copper in Descemet's membrane of the cornea); blue lunulae of the fingernails; and high degree of copper storage in the body.

C. HFE-Associated Hereditary Hemochromatosis (HHH).
1. HHH is an autosomal recessive genetic disorder caused by ≈28 different mutations in the *HFE* gene on **chromosome 6p21.3** for **hereditary hemochromatosis protein,** which is a cell surface protein, expressed as a heterodimer with ß$_2$-microglobulin, binds the transferrin receptor 1, and reduces cellular iron uptake although the exact mechanism is unknown.
2. HHH is most commonly caused by two missense mutations that result in a **normal cysteine → tyrosine** substitution at position 282 (C282Y), resulting in decreased cell surface expression or that result in a **normal histidine → asparagine** substitution at position 63 (H63D), resulting in pH changes that affects binding to the transferrin receptor 1.
3. ≈87% of HHH affected individuals in the European population are homozygous for the C282Y mutation or are **compound heterozygous** (i.e., two different mutations at the same gene locus) for the C282Y and H63D mutations.
4. These mutations result in elevated transferrin-iron saturation, elevated serum ferritin concentration, and hepatic iron overload assessed by **Prussian blue** staining of a liver biopsy.
5. If a person has HHH decides to have a child, then the carrier risk factor becomes important. The risk that a partner of European descent is a heterozygote (Hh) is 11% (1 out of 9 individuals), due the high carrier rate in the general European population for HHH.
6. **Prevalence.** The prevalence of HHH is 1/200 to 500 births.
7. **Clinical features include:** excessive storage of iron in the liver, heart, skin, pancreas, joints, and testes; abdominal pain, weakness, lethargy, weight loss, and hepatic fibrosis; without therapy, symptoms appear in males at 40 to 60 years of age and in females after menopause.

VII. METABOLIC GENETIC DISORDERS INVOLVING DEGRADATION PATHWAYS

Most complex biomolecules are recycled by degradation into simpler molecules, which can then be eliminated or used to synthesize new molecules. Malfunctions in degradation pathways will result in the accumulation (or "storage") of complex biomolecules within the cell. For example, lysosomal enzymes catalyze the stepwise degradation of glycosaminoglycans (GAGs; formerly called mucopolysaccharides), sphingolipids, glycoproteins, and glycolipids. **Lysosomal storage disorders (or mucopolysaccharidoses)** are caused by lysosomal enzyme deficiencies required for the stepwise degradation of GAGs that result in the accumulation of partially degraded GAGs within the cell, leading to organ dysfunction.

A. Mucopolysaccharidosis Type I (MPS I).

1. MPS I is an autosomal recessive genetic disorder caused by ≈57 different mutations in the **IDUA gene** on **chromosome 4p16.3** for **α-L-iduronidase** that catalyzes the reaction that removes α-L-iduronate residues from heparan sulfate and dermatan sulphate during lysosomal degradation.

2. MPS I is the prototypical mucopolysaccharidoses disorder. MPS I presents as a continuum from severe to mild clinical symptoms, and MPS I affected individuals are best described as having either **severe symptoms (MPS IH; Hurler syndrome); intermediate symptoms (MPS IH/S; Hurler-Scheie syndrome); or mild symptoms (MPS IS; Scheie syndrome).**

3. MPS IH (Hurler syndrome) is most commonly caused by two nonsense mutations which result in a **normal tryptophan → nonsense** substitution at position 402 (W402X) or in a **normal glutamine → nonsense** substitution at position 70 (Q70X).

4. The W402X mutation accounts for 55% of cases in the Australasian population. The Q70X mutation accounts for 65% of cases in the Scandinavian population.

5. These mutations result in elevated heparan sulphate and dermatan sulphate excretion in the urine, reduced/absent α-L-iduronidase activity, and **heparan sulfate** and **dermatan sulfate** accumulation.

6. Prevalence. The prevalence of MPS IH is 1/100,000 and of MPS IS is 1/500,000.

7. Clinical features of MPS IH (Hurler syndrome) include: infants initially appear normal up to ≈9 months of age but then develop symptoms; coarsening of facial features, thickening of alae nasi, lips, ear lobules, and tongue; corneal clouding; severe visual impairment; progressive thickening heart valves leading to mitral and aortic regurgitation; dorsolumbar kyphosis; skeletal dysplasia involving all the bones; linear growth ceases by 3 years of age; hearing loss; chronic recurrent rhinitis; severe mental retardation; and zebra bodies within neurons.

B. Gaucher Disease (GD).

1. GD is an autosomal recessive genetic disorder caused by mutations in the **GBA gene** on **chromosome 1q21** for **β-glucosylceramidase,** which hydrolyzes glucocerebroside into glucose and ceramide.

2. GD is the most common lysosomal storage disorder. GD presents as a continuum of clinical symptoms and is divided into three major clinical types **(Types 1, 2, and 3)** which is useful in determining prognosis and management of the individual.

3. GD is most commonly caused by either a missense mutation which results in a **normal asparagine → serine** substitution at position 370 (N370S), a missense mutation which results in a **normal leucine → proline** substitution at position 444 (L444P), a 84GG mutation, or a IVS2+1 mutation.

4. The N370S, L444P, 84GG, and IVS2+1 mutations account for 95% of cases in the Ashkenazi Jewish population. These mutations result in absent/near absent ß-glucosylceramidase activity and **glucosylceramide (and other glycolipids)** accumulation.

5. If one parent has GD (gg), the risk that a partner of Ashkenazi Jewish descent is a heterozygote is ≈5% (1 out of 18 individuals) due the high carrier rate in the general Ashkenazi Jewish population.

6. **Prevalence.** The prevalence of Type I GD is 1/855 in the Ashkenazi Jewish population.
7. **Clinical features of Type I GD include:** bone disease (e.g., focal lytic lesions, sclerotic lesions, osteonecrosis) is the most debilitating pathology of Type I GD; hepatomegaly; splenomegaly; cytopenia and anemia due to hypersplenism, splenic sequestration, and decreased erythropoiesis; and pulmonary disease (e.g., interstitial lung disease, alveolar/lobar consolidation; pulmonary hypertension); no primary CNS involvement.

C. **Hexosaminidase A Deficiency (HAD).**
   1. **Acute infantile HAD (Tay-Sachs disease; TSD)** is the prototypical HAD. HAD presents as a group of neurodegenerative disorders caused by lysosomal accumulation of **GM2 ganglioside.**
   2. TSD is an autosomal recessive genetic disorder caused by mutations in the **HEXA gene** on **chromosome 15q23-q24** for **hexosaminidase α-subunit,** which catalyzes the reaction that cleaves the terminal ß-linked N-acetylgalactosamine from GM2 ganglioside.
   3. TSD is most commonly caused by either a **4-bp insertion in exon 11 mutation** (+TATC1278) which produces a frameshift and a premature STOP codon or a **RNA splicing mutation in intron 12** (+1IVS12) which produces unstable mRNAs, which are probably rapidly degraded.
   4. The +TATC1278 and the +1IVS12 mutations account for ≈95% of cases in the Ashkenazi Jewish population. These mutations result in absent/near absent hexosaminidase A activity and **GM2 ganglioside** accumulation.
   5. **Prevalence.** The prevalence of TSD is 1/324,000 births in the Ashkenazi Jewish population since the advent of population-based carrier screening. The prevalence of TSD was 1/3,600 births in the Ashkenazi Jewish population before the advent of population-based carrier screening.
   6. **Clinical features of TSD include:** infants initially appear normal up to 3 to 6 months of age but then develop symptoms; progressive weakness and loss of motor skills; decreased attentiveness; increased startled response; a **cherry red spot in the fovea centralis** of the retina; generalized muscular hypotonia; later, progressive neurodegeneration, seizures, blindness, and spasticity occur followed by death at ≈2 to 4 years of age.

D. **Other Genetic Disorders Involving Degradation Pathways.** These include: mucopolysaccharidosis type II (MPS II; Hunter syndrome); mucopolysaccharidosis type IIIA (MPS IIIA; Sanfilippo A syndrome); mucopolysaccharidosis type IVA (MPS IVA; Morquio A syndrome); Niemann-Pick (NP) type 1A disorder; Fabry disorder; Krabbe disorder; and metachromatic leukodystrophy (MLD).

VIII. SUMMARY TABLES OF METABOLIC GENETIC DISORDERS
(Tables 12-1, 12-2, 12-3, 12-4, 12-5, and 12-6)

IX. SELECTED PHOTOGRAPHS OF METABOLIC GENETIC DISORDERS (Figure 12-1)

t a b l e **12-1**	Metabolic Genetic Disorders Involving Carbohydrate Pathways

Genetic Disorder	Gene/Gene Product Chromosome	Clinical Features
Galactosemia	*GALT* gene/galactose-1-phosphate uridylyltransferase 9p13	Feeding problems in the newborn, failure to thrive, hypoglycemia, hepatocellular damage, bleeding diathesis, jaundice, and hyperammonemia; sepsis with *E. coli*, shock, and death may occur if the galactosemia is not treated
Asymptomatic fructosuria	*KHK* gene/ketohexokinase or fructokinase 2p23.3-23.2	Presence of fructose in the urine
Hereditary fructose intolerance	*ALDOB* gene/fructose 1-phosphate aldolase B 9q21.3-q22.2	Failure to thrive, fructosuria, hepatomegaly, jaundice, aminoaciduria, metabolic acidosis, lactic acidosis, low urine ketones, recurrent hypoglycemia and vomiting at the age of weaning when fructose or sucrose (a disaccharide that is hydrolyzed to glucose and fructose) is added to the diet, and infant and adults are asymptomatic until they ingest fructose or sucrose
Lactose intolerance	*LCT* gene/lactase-phlorizin hydrolase 2q21	Diarrhea, crampy abdominal pain localized to the periumbilical area or lower quadrant, flatulence, nausea, vomiting, audible Borborygmi, stools that are bulky, frothy, and watery, and bloating after milk or lactose consumption
GSD type Ia; von Gierke

GSD type Ib; von Gierke | *G6PC* gene/glucose-6-phosphatase 17q21
SLC37A4 gene/glucose-6-phosphate translocase 11q23 | Accumulation of glycogen and fat in the liver and kidney resulting in hepatomegaly and renomegaly, severe hypoglycemia, lactic acidosis, hyperuricemia, hyperlipidemia, hypoglycemic seizures, doll-like faces with fat cheeks, relatively thin extremities, short stature, protuberant abdomen, and neutropenia with recurrent bacterial infections. |
GSD type V; McArdle	*PYGM* gene/muscle glycogen phosphorylase 11q13	Exercise-induced muscle cramps and pain, "second wind" phenomenon with relief of myalgia and fatigue after a few minutes of rest, episodes of myoglobinuria, increased resting basal serum creatine kinase (CK) activity, onset typically occurs around 20–30 years of age; clumsiness, lethargy, slow movement, and laziness in preadolescents.
GSD type II; Pompe	*GAA* gene/lysosomal acid a-glucosidase 17q25.2-q25.3	Muscle and heart are affected
GSD type IIIa; Cori	*AGL* gene/amylo-1,6glucosidase, 4-a-glucanotransferase (or glycogen branching enzyme) 1p21	Muscle and liver are affected
GSD type IV; Andersen	*GBE1* gene/glucan(1,4-a-) branching enzyme 1 (or glycogen branching enzyme) 3	Muscle and liver are affected
GSD type VI; Hers	*PYGL* gene/liver glycogen phosphorylase 14q11.2-q24.3	Liver is affected
GSD type VII; Tarui	*PFKM* gene/muscle phosphofructokinase 12q13.11	Muscle is affected

GSD, glycogen storage disease.

t a b l e **12-2** Metabolic Genetic Disorders Involving Amino Acid Pathways		
Genetic Disorder	**Gene/Gene Product Chromosome**	**Clinical Features**
Phenylalanine hydrolase deficiency (classic phenylketonuria [PKU])	*PAH* gene/phenylalanine hydrolase 12q23.2	No physical signs are apparent in neonates with PAH deficiency; diagnosis is based on detection of elevated plasma PAH concentration (>1,000 umol/L for classic PKU) and normal BH$_4$ cofactor metabolism; a dietary phenylalanine tolerance of <500 mg/day; untreated children with classic PKU show impaired brain development, microcephaly, epilepsy, severe mental retardation, behavioral problems, depression, anxiety, musty body odor, and skin conditions like eczema.
Hereditary tyrosinemia type I	*FAH* gene/fumarylacetoacetate hydrolase 15q23-q25	Diagnosis is based on detection of elevated plasma succinylacetone concentration, elevated plasma tyrosine, methionine, and phenylalanine concentrations, elevated urinary tyrosine metabolite (e.g., hydroxyphenylpyruvate) concentration, elevated urinary δ-aminolevulinic acid; cabbagelike odor; untreated children with HTI show sever liver dysfunction, renal tubular dysfunction, growth failure, and rickets.
Maple syrup urine disease	*BCKDHA* gene/E1a subunit of branched-chain ketoacid dehydrogenase complex (BCKD) 19q13.1-q13.2 *BCKDHB* gene/E1ß subunit of BCKD 6q14 *DBT* gene/E2 subunit of BCKD 1p31	Untreated children with MSUD show maple syrup odor in cerumen 12–24 hours after birth, elevated plasma branched-chain amino acid concentration, ketonuria, irritability, poor feeding by 2–3 days of age, deepening encephalopathy including lethargy, intermittent apnea, opisthotonus, and stereotyped movements like "fencing" and "bicycling" by 4–5 days of age; acute leucine intoxication (leucinosis) associated with neurological deterioration due to the ability of leucine to interfere with the transport of other large neutral amino acids across the blood–brain barrier thereby reducing the amino acid supply to the brain.

t a b l e **12-3** Metabolic Genetic Disorders Involving Lipid Pathways		
Genetic Disorder	**Gene/Gene Product Chromosome**	**Clinical Features**
Medium-chain acyl-coenzyme A dehydrogenase deficiency	*ACADM* gene/medium-chain acyl-coenzyme A dehydrogenase 1p31	Hyperketotic hypoglycemia, vomiting, and lethargy triggered by either a common illness (e.g., viral gastrointestinal or upper respiratory tract infections) or prolonged fasting (e.g., weaning the infant from nighttime feedings) which may quickly progress to coma and death; hepatomegaly and acute liver disease; children are normal at birth and present between 3 and 24 months of age; later presentation into adulthood is possible.
Smith-Lemli-Opitz syndrome	*DHCR7* gene/7-dehydrocholesterol reductase 11q12-q13	Prenatal and postnatal growth retardation, microcephaly, moderate to severe mental retardation, cleft palate, cardiac defects, underdeveloped external genitalia and hypospadias in males, postaxial polydactyly, Y-shaped 2-3 toe syndactyly, downslanting palpebral fissures, epicanthal folds, anteverted nares, and micrognathia.
Familial hypercholesterolemia	*LDLR* gene/low-density lipoprotein receptor 19p13.1-13.3 *APOB* gene/apolipoprotein B-100 2p23-p24 *PCSK9* gene/proprotein convertase subtilisin/kexin type 9 1p32-34.1 *PCSK9* gene Tyr142Stop, Cys679Stop, Arg46Leu mutations	Premature heart disease as a result of atheromas (deposits of LDL-derived cholesterol in the coronary arteries), xanthomas (cholesterol deposits in the skin and tendons), arcus lipoides (deposits of cholesterol around the cornea of the eye), homozygote and heterozygote phenotypes are known, homozygotes develop severe symptoms early in life and rarely live past 30 years of age, heterozygotes have plasma cholesterol level twice that of normal. Hypocholesterolemia

t a b l e **12-4**	Metabolic Genetic Disorders Involving the Urea Cycle Pathway	
Genetic Disorder	**Gene/Gene Product Chromosome**	**Clinical Features**
Ornithine transcarbamylase deficiency	OTC gene/ornithine transcarbamylase Xp21.1	Infants initially appear normal but then rapidly develop hyperammonemia, cerebral edema, lethargy, anorexia, hyperventilation or hypoventilation, hypothermia, seizures, neurologic posturing, and coma; in infants with partial enzyme deficiencies, the symptoms may be delayed for months or years, the symptoms are more subtle, the hyperammonemia is less severe, and ammonia accumulation can be triggered by illness or stress throughout life.
Carbamoylphosphate synthetase I deficiency	CPS1 gene/carbamoylphosphate synthetase 1 2q35	
Argininosuccinic acid synthetase deficiency (or citrullinemia type I)	ASS gene/argininosuccinic acid synthetase 9q34	
Argininosuccinic acid lyase deficiency (or argininosuccinic aciduria)	ASL gene/argininosuccinic acid lyase 7cen-q11.2	OTC deficiency and CPSI deficiency are the most severe types of urea cycle disorders.
Arginase deficiency (or hyperargininemia)	ARG1 gene/arginase 6q23	
N-acetyl glutamine synthetase deficiency	NAGS gene/N-acetyl glutamate synthetase 17q21.3	

t a b l e **12-5**	Metabolic Genetic Disorders Involving Transport Pathways	
Genetic Disorder	**Gene/Gene Product Chromosome**	**Clinical Features**
Menkes disease	ATP7A gene/copper-transporting ATPase 1 Xq12-q1	Infants initially appear normal up to 2–3 months of age but then develop hypotonia, seizures, failure to thrive, loss of developmental milestones, changes in hair (short, coarse, twisted, lightly pigmented, "steel wool" appearance), jowly facial appearance with sagging cheeks, temperature instability, hypoglycemia, urinary bladder diverticula, and gastric polyps. Without early treatment with parenteral copper, MND progresses to severe neurodegeneration and death by 7 months → 3 years of age.
Wilson disease	ATP7B gene/copper transporting ATPase 2 13q14.3-q21.1	Symptoms occur in individuals from 3–50 years of age, recurrent jaundice, hepatitislike illness, fulminant hepatic failure, tremors, poor coordination, loss of fine motor control, chorea, masklike facies, rigidity, gait disturbance, depression, neurotic behaviors, Kayser-Fleischer rings (deposition of copper in Descemet membrane of the cornea), blue lunulae of the fingernails, and high degree of copper storage in the body.
HFE-associated hereditary hemochromatosis	HFE gene/hereditary hemochromatosis protein 6p21.3	Excessive storage of iron in the liver, heart, skin, pancreas, joints, and testes; abdominal pain, weakness, lethargy, weight loss, and hepatic fibrosis; without therapy, symptoms appear in males at 40–60 years of age and in females after menopause.

| table | **12-6** | Metabolic Genetic Disorders Involving Degradation Pathways |

Genetic Disorder	Gene/Gene Product Chromosome	Accumulation Product	Clinical Features
Mucopolysaccharidosis type I (Hurler , Hurler-Scheie, or Scheie syndromes)	*IDUA* gene/a-L-iduronidase 4p16.3	Heparan sulfate Dermatan sulfate	Infants initially appear normal up to ≈9 months of age but then develop symptoms; coarsening of facial features, thickening of alae nasi, lips, ear lobules, and tongue; corneal clouding; severe visual impairment; progressive thickening heart valves leading to mitral and aortic regurgitation; dorsolumbar kyphosis; skeletal dysplasia involving all the bones; linear growth ceases by 3 years of age; hearing loss; chronic recurrent rhinitis; severe mental retardation; and zebra bodies within neurons.
Gaucher disease	*GBA* gene/β-glucosylceramidase 1p21	Glucosylceramide Other glycolipids	Bone disease (e.g., focal lytic lesions, sclerotic lesions, osteonecrosis) is the most debilitating pathology of Type I GD; hepatomegaly; splenomegaly; cytopenia and anemia due to hypersplenism, splenic sequestration, and decreased erythropoiesis; and pulmonary disease (e.g., interstitial lung disease, alveolar/lobar consolidation; pulmonary hypertension); no primary CNS involvement.
Hexosaminidase A deficiency (Tay-Sachs)	*HEXA* gene/hexosaminidase a-subunit 15q23-q24	GM2 ganglioside	Infants initially appear normal up to 3–6 months of age but then develop symptoms; progressive weakness and loss of motor skills; decreased attentiveness; increased startled response; a cherry red spot in the fovea centralis of the retina; generalized muscular hypotonia; later, progressive neurodegeneration, seizures, blindness, and spasticity occur followed by death at ≈2–4 years of age.
Mucopolysaccharidosis type II (Hunter syndrome)	*IDS* gene/iduronate 2-sulfatase Xq27.3-q28	Heparan sulfate Dermatan sulfate	Dysostosis multiplex (thickened skull, anterior thickening of ribs, vertebral abnormalities, and short/thick long bones), coarse face, hepatosplenomegaly, mental retardation, and behavioral problems.
Mucopolysaccharidosis type III A (Sanfilippo A syndrome)	*SGSH* gene/sulfatidase 17q25.3	Heparan sulfate	Dysostosis multiplex (thickened skull, anterior thickening of ribs, vertebral abnormalities, and short/thick long bones), mental retardation, and behavioral problems (aggressive behavior followed by progressive neurological decline).
Mucopolysaccharidosis type IV A (Morquio A syndrome)	*GALNS* gene/N-acetylgalactosamine-6-sulfate sulfatase 16q24.3	Keratan sulfate Chondroitin-6-sulfate	Short stature, bony dysplasia, and hearing loss.
Niemann-Pick type 1A disorder	*SMPD1* gene/acid sphingomyelinase 11p15	Sphingomyelin	Hepatosplenomegaly, feeding difficulties, and loss of motor skills is seen within 1–3 months of age; a cherry red spot in the fovea centralis of the retina; later, a rapid, profound, and progressive neurodegeneration occurs followed by death at ≈2–3 years of age.

(continued)

table **12-6** *(continued)*

Genetic Disorder	Gene/Gene Product Chromosome	Accumulation Product	Clinical Features
Fabry disorder	*GLA* gene/a-galactosidase A Xq22	Globotriaosyl-ceramide	In classically affected males, symptoms begin in childhood or adolescence; severe neuropathic or limb pain precipitated by stress, extreme heat or cold, and physical exertion; telangiectasias, angiokeratomas in the groin, hip, and periumbilical regions; asymptomatic corneal deposits; retinal vascular tortuosity; in adulthood, end stage renal disease, cardiac involvement, and cerebrovascular involvement leading transient ischemic attacks and strokes.
Krabbe disorder	*GALC* gene/ galactocerebrosidase 14q31	Galactosylceramide	Developmental delay, limb stiffness, hypotonia, absent reflexes, optic atrophy, microcephaly, and extreme irritability within 1–6 months of age; later, seizures and tonic extensor spasms associated with light, sound, or touch stimulation occur; a rapid regression to the decerebrate condition followed by death at ≈2 years of age.
Metachromatic leukodystrophy	*ARSA* gene/ arylsulfatase A 22q13.3-qter	Cerebroside sulfate	Regression of motor skills, gait difficulties, ataxia, hypotonia, extensor plantar responses, and optic atrophy within 6 months to 2 years of age; later, peripheral neuropathy occurs followed by death at ≈5–6 years of age; metachromatic lipid deposits in neurons is pathognomic

FIGURE 12-1. (A1, A2, A3) Glycogen storage disease type I (von Gierke). (A1) Photograph shows a boy 13 years of age with von Gierke disease. Note the enlarged abdomen. **(A2)** Light micrograph of a liver biopsy shows hepatocytes with a pale, clear cytoplasm due to the large amounts of accumulated glycogen that is extracted during histological processing. **(A3)** Electron micrograph of a liver biopsy shows hepatocytes filled with glycogen aggregates. **(B) Glycogen storage disease type V (McArdle).** Light micrograph of a skeletal muscle biopsy shows muscle cells with a pale, clear cytoplasm due to the large amounts of accumulated glycogen that is extracted during histological processing. **(C) Maple syrup urine disease.** T2-weighted MRI shows hyperdensity in the brain stem (arrows) indicating neurological deterioration. **(D1, D2) Familial hypercholesterolemia (D1)** Photograph shows xanthomas on the dorsum of the hand. **(D2)** Photograph shows arcus lipoides, which represents the deposition of cholesterol around the cornea of the eye. **(E) Wilson disease.** Photograph shows a Kayser-Fleischer ring (arrows) caused by the deposition of copper in Descemet membrane and thereby obstructs the view of the underlying iris. **(F) Hemochromatosis.** Light micrograph of a liver biopsy stained with Prussian blue shows hepatocytes with a heavily stained cytoplasm to the large amounts of accumulated iron. **(G) Mucopolysaccharidosis type I (MPS I; Hurler syndrome).** Photograph shows an infant with coarsening of facial features; thickening of alae nasi, lips, ear lobules, and tongue; dorsolumbar kyphosis; and skeletal dysplasia involving all the bones. **(H) Gaucher disease.** Light micrograph of a bone marrow aspirate smear shows the typical Gaucher cells with an abundant cytoplasm filled with fibrillarlike material. **(I) Tay-Sachs disease.** Electron micrograph shows central nervous system neurons filled lysosomes (Lys) containing whorls of membranelike material due to the large amounts of accumulated GM2 ganglioside. N = nucleus of neuron.

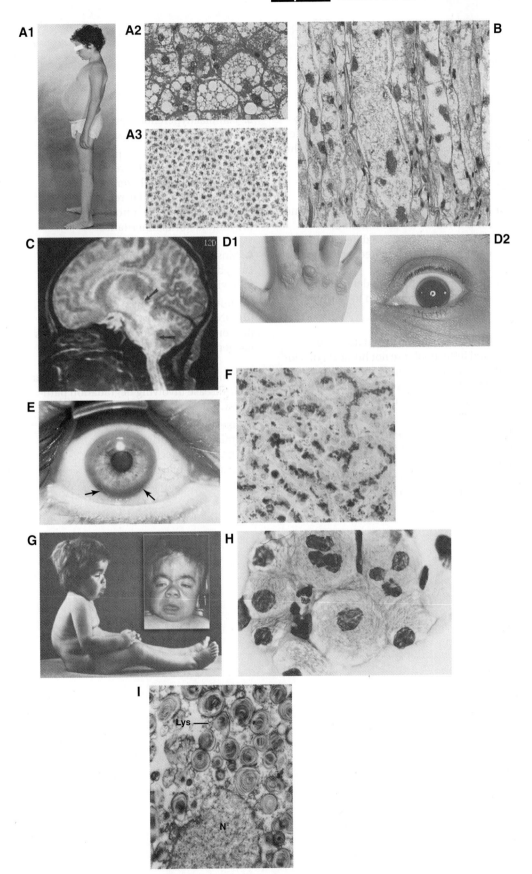

Review Test

1. In metabolic genetic diseases, heterozygotes are generally normal. Which of the following is the most likely explanation for this observation?

(A) Heterozygotes have compound mutations.

(B) Heterozygotes produce enough enzymes for normal metabolic function.

(C) Heterozygotes produce defective enzymes that are repaired by the normal allele.

(D) Heterozygotes produce defective enzymes that are spliced to normal enzymes.

2. In hereditary hemochromatosis, some affected individuals are not homozygous for either of the two common mutations, C282Y and H63D, which cause the disease. Which one of the following causes the disease in these individuals?

(A) high hepatic concentration of copper

(B) heterozygosity for C282Y

(C) heterozygosity for H63D

(D) compound heterozygosity for the two common mutations

3. Familial hypercholesterolemia is caused by which one of the following?

(A) autosomal recessive inheritance of a mutation in the LDL receptor gene

(B) homozygosity or heterozygosity for a mutation in the LDL receptor gene

(C) a mutation in the mitochondria

(D) multifactorial inheritance

4. Most genetic metabolic diseases are caused by mutations in which one of the following?

(A) mitochondrial genes

(B) DNA repair genes

(C) genes that code for structural proteins

(D) genes that code for enzymes

5. Which of the following is a genetic metabolic disease involving degradation pathways?

(A) glycogen storage disease

(B) maple syrup urine disease

(C) mucopolysaccharidosis

(D) galactosemia

Answers and Explanations

1. **The answer is (B).** Because most metabolic diseases are inherited in an autosomal recessive fashion, heterozygotes have a normal allele and usually produce enough enzymes for normal metabolic function.

2. **The answer is (D).** Heterozygotes for either C282Y or H63D are normal, but when both mutations are present in the heterozygous state in an individual that individual is affected. This is called compound heterozygosity.

3. **The answer is (B).** Familial hypercholesterolemia is caused by autosomal dominant inheritance of a mutation in the low density lipoprotein receptor (LDLR) gene. Because it is an autosomal dominant disease, both heterozygous and homozygous individuals are affected. Most cases of hypercholesterolemia in the population are multifactorial in origin.

4. **The answer is (D).** Metabolic reactions within biochemical pathways require enzymes to facilitate the reactions. Most metabolic diseases are caused by mutations in the genes that encode the enzymes in a biochemical pathway.

5. **The answer is (C).** Mucopolysaccharidosis is due to lysosomal enzyme deficiencies that result in the lack of complete degradation of glycosaminoglycans, resulting in the buildup of incomplete degradation products in cells.

13 Genetics of Hemoglobinopathies

I. CHARACTERISTICS OF HEMOGLOBIN (Hb)

A. Hb is a globular protein consisting of **four subunits.**

B. **Fetal Hb (HbF)** consists of two alpha-globin subunits and two gamma-globin subunits designated **Hb $\alpha_2\gamma_2$.** HbF is the **major form of Hb during fetal development** because the O_2 affinity of HbF is higher than the O_2 affinity of HbA and thereby "pulls" O_2 from the maternal blood into fetal blood.

C. **Adult Hb (HbA)** consists of two alpha-globin subunits and two beta-globin subunits designated **Hb $\alpha_2\beta_2$.**

D. Hb contains a **heme moiety**, which is an **iron (Fe)-containing porphyrin.** Fe^{2+} (ferrous state) binds O_2, forming **oxyhemoglobin.** Fe^{3+} (ferric state) does not bind O_2, forming **deoxyhemoglobin.** The heme moiety is synthesized partially in mitochondria and partially in cytoplasm.

II. SICKLE CELL DISEASE (SCD)

A. SCD is an autosomal recessive genetic disorder caused by a missense mutation ($GAG \rightarrow GTG$) at the second nucleotide of the sixth codon in the **HBB** gene on **chromosome 11p15.5** which results in a normal **glutamic acid → valine** substitution **(E6V)** in the β-globin subunit of **hemoglobin.**

B. SCD is defined by the presence of **E6V HbS** and accounts for 60% to 70% of SCD cases in the United States.

C. E6V HbS forms highly ordered polymers that aggregate and distort the shape of red blood cells, making them brittle and poorly deformable.

D. SCD may also be caused by coinheritance of the E6V *HBB* gene mutation with another *HBB* gene mutation as follows:
1. E6V HBB gene mutation + a missense mutation, which results in a normal **glutamic acid → lysine** substitution **(E6K)** in the *HBB* gene forming **HbC.** This means that both **E6V HbS and E6K HbC** will be present within red blood cells.
2. E6V HBB gene mutation + various *HBB* gene mutations associated with β-thalassemia. This means that both **E6V HbS and various mutations of Hb** will be present within red blood cells.

3. E6V HBB gene mutation + a missense mutation, which results in a normal **glutamic acid → glutamine** substitution (**E121Q**) in the *HBB* gene forming **Hb D (D-Punjab)**. This means that both **E6V HbS and E121Q HbD** will be present within red blood cells.
4. E6V HBB gene mutation + a missense mutation, which results in a normal **glutamic acid → lysine** substitution (**E121K**) forming **Hb O (O-Arab)**. This means that both **E6V HbS and E121K HbO** will be present within red blood cells.

E. Sickle cell disease has a very high carrier frequency in African, Mediterranean, Middle Eastern, Indian, Caribbean, and portions of Central and South American populations. In particular, sickle cell disease has a 1 in 12 carrier frequency in the African American population and a 1 in 4 carrier frequency in the west central African population.

F. **Prevalence.** The prevalence of SCD 1/200 to 650 births in the African American population.

G. **Clinical features include:** infants appear healthy at birth but become symptomatic later after fetal hemoglobin (HbF) levels decrease and HbS levels increase (note: HbF does not contain B-globin subunits); pain and/or swelling of hands and feet in infants and young children; varying degrees of hemolysis leading to chronic anemia, cholelithiasis, and delayed growth and sexual maturation; intermittent episodes of vascular occlusion; in patients with osteomyelitis there is a disproportionate number of cases due to *Salmonella* infection; functional asplenia usually results in adolescence after so-called autoinfarction of the spleen; and acute and chronic organ dysfunction.

III. ALPHA-THALASSEMIA

A. α-Thalassemia is an autosomal recessive genetic disorder most commonly caused by a deletion of the *HBA1 gene* and/or the *HBA2 gene* on **chromosome 16pter-p13.3** for the α_1–globin subunit of *hemoglobin* and α_2-globin subunit of *hemoglobin*, respectively. These deletions occur during unequal crossing over between homologous chromosomes during meiosis.

B. α-Thalassemia is defined by the reduced synthesis of α-globin subunits of hemoglobin. It should be noted that the clinical amount of α-globin subunits of hemoglobin is due to **four (4) alleles.**

C. The deletions of the *HBA1 gene* and/or the *HBA2 gene* result in the **reduced amounts of HbF (Hb $\alpha_2\gamma_2$)** and **HbA (Hb $\alpha_2\beta_2$)** because there is reduced synthesis of α-globin subunits, which are common to both HbF and HbA.

D. There are $\approx$45 nondeletional mutations that cause α-thalassemia. The most common nondeletional mutation is a mutation in the STOP codon of the *HBA2* gene resulting in an α-globin subunit elongated by 31amino acids called **Hb**$^{\text{Constant Spring}}$.

E. α-Thalassemia has two carrier states:
1. α^0**-Thalassemia (or α-thalassemia trait)**
   a. α^0-thalassemia results from the inheritance of a deletion or dysfunction of two α-globin alleles. This is a **carrier state** for α-thalassemia.
   b. **Clinical features include:** moderate thalassemialike hematologic findings.
2. α^+**-Thalassemia**
   a. α^+-Thalassemia results from the inheritance of a deletion or dysfunction of one α-globin allele. This is a **silent carrier state** for α-thalassemia.
   b. **Clinical features include:** normal hematologic findings or moderate thalassemialike hematologic findings.

F. **Prevalence.** The prevalence of α-thalassemia is very high in the African, Mediterranean, Arabic, Indian, and Southeast Asian populations. The prevalence of Hb Bart hydrops fetalis

$\approx 1/200$ births in India. The prevalence of HbH disease is $\approx 1/50$ births in India. The prevalence of α^+-thalassemia (silent carriers) is $1/6$ births in Sardinia.

G. There are two clinically significant forms of α-thalassemia:
  1. **Hb Bart hydrops fetalis syndrome (Hb Bart).**
     a. **Hb Bart** results from the inheritance of a deletion or dysfunction of all four α-globin alleles and is the most severe form of α-thalassemia.
     b. An **excess of γ-globin subunits** form tetramers during fetal development that have extremely high affinity for oxygen but are unable to deliver oxygen to fetal tissues.
     c. **Clinical features include:** fetal onset of generalized edema, ascites, pleural and pericardial effusions, severe hypochromatic anemia, and death in the neonatal period.
  2. **Hemoglobin H (HbH) disease.**
     a. HbH results from the inheritance of a deletion or dysfunction of three α-globin alleles.
     b. A relative **excess of β-globin subunits** form insoluble inclusion bodies within mature red blood cells.
     c. **Clinical features include:** mild microcytic hypochromatic hemolytic anemia and hepatosplenomegaly.

IV. BETA-THALASSEMIA

A. β-Thalassemia is an autosomal recessive genetic disorder caused by >200 missense or frameshift mutations in the **HBB gene** on **chromosome 11p15.5** for the **β-globin subunit of hemoglobin.**

B. β-Thalassemia is defined by the absence or reduced synthesis of β-globin subunits of hemoglobin. A **β^0 mutation** refers to a mutation that causes the absence of β-globin subunits. A **β^+ mutation** refers to a mutation that causes the reduced synthesis of β-globin subunits. It should be noted that the clinical amount of β-globin subunits of hemoglobin is due to **two (2) alleles.**

C. The mutations in the **HBB gene** result in the **reduced amounts of HbA (Hb $\alpha_2\beta_2$)** because there is reduced synthesis of β-globin subunits, which are found only in HbA.

D. Heterozygote carriers of β-thalassemia are often referred to as having **thalassemia minor.**
  1. Thalassemia minor results from the inheritance of a β^+ mutation of one β-globin allele (β^+/normal β).
  2. **Clinical features include:** individuals are asymptomatic with very mild or absent anemia, but red blood cell abnormalities may be seen.

E. Prevalence. The prevalence of β-thalassemia is very high in the African, Mediterranean, Arabic, Indian, and Southeast Asian populations. The prevalence of β-thalassemia is $1/7$ births in Cyprus and $1/8$ births in Sardinia.

F. There are two clinically significant forms of β-thalassemia:
  1. **Thalassemia major.**
     a. Thalassemia major results from the inheritance of a β^0 mutation of both β-globin alleles (β^0/β^0) and is the most severe form of β-thalassemia.
     b. An **excess of α-globin subunits** form insoluble inclusion bodies within mature red blood cell precursors.
     c. **Clinical features include:** microcytic hypochromatic hemolytic anemia, abnormal peripheral blood smear with nucleated red blood cells, reduced amounts of HbA, severe anemia, hepatosplenomegaly, fail to thrive, become progressively pale, regular blood transfusion are necessary, and usually come to medical attention between 6 months → 2 years of age.

2. Thalassemia intermedia.

    a. Thalassemia intermedia results from the inheritance of a β^0 mutation of one β-globin allele (β^0/normal β) and is a less severe form of β-thalassemia.

    b. Clinical features include: a mild hemolytic anemia, individuals are at risk for iron overload, regular blood transfusions are rarely necessary, and usually come to medical attention by >2 years of age.

V. SUMMARY TABLE OF HEMOGLOBINOPATHIES (Table 13-1)

t a b l e   13-1   Summary Table of Hemoglobinopathies

Genetic Disorder	Gene/Gene Product Chromosome	Clinical Features
Sickle cell disease	*HBB* gene/β-globin subunit 11p15.5 glutamic acid → valine missense mutation very common (E6V)	Infants appear healthy at birth but become symptomatic later after fetal hemoglobin (HbF) levels decrease and HbS levels increase; pain and/or swelling of hands and feet in infants and young children; varying degrees of hemolysis leading to chronic anemia, cholelithiasis, and delayed growth and sexual maturation; intermittent episodes of vascular occlusion; and acute and chronic organ dysfunction.
α-Thalassemia	*HBA1* gene/α_1-globin subunit *HBA2* gene/α_2-globin subunit 16pter-p13.3 Deletion mutation is most common Reduced amounts of HbF (Hb $\alpha_2\gamma_2$) and HbA (Hb $\alpha_2\beta_2$)	**Hb Bart**: fetal onset of generalized edema, ascites, pleural and pericardial effusions, severe hypochromatic anemia, and death in the neonatal period. **HbH**: mild microcytic hypochromatic hemolytic anemia and hepatosplenomegaly. α^0-**Thalassemia (or α-thalassemia trait; carrier state):** moderate thalassemialike hematologic findings. α^+-**Thalassemia (silent carrier state):** normal hematologic findings or moderate thalassemialike hematologic findings.
β-Thalassemia	*HBB* gene/β-globin subunit 11p15.5 >200 missense or frameshift mutations are most common Reduced amounts of HbA (Hb $\alpha_2\beta_2$)	**Thalassemia major (β^0/β^0):** microcytic hypochromatic hemolytic anemia, abnormal peripheral blood smear with nucleated red blood cells, reduced amounts of HbA, severe anemia, hepatosplenomegaly, fail to thrive, become progressively pale, regular blood transfusion are necessary, and usually come to medical attention between 6 months and 2 years of age. **Thalassemia intermedia (β^0/normal β):** a mild hemolytic anemia, individuals are at risk for iron overload, regular blood transfusions are rarely necessary, and usually come to medical attention by >2 years of age. **Thalassemia minor (carrier state; β^+/normal β):** individuals are asymptomatic with very mild or absent anemia, but red blood cell abnormalities may be seen.

VI. SELECTED PHOTOMICROGRAPHS OF HEMOGLOBINOPATHIES (Figure 13-1)

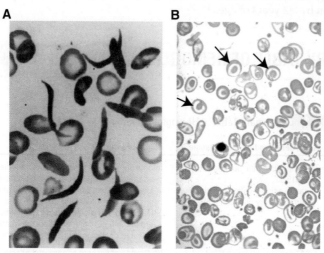

A   **B**

Figure 13-1. Hemoglobinopathies. (A) Sickle cell disease. Photomicrograph shows sickle RBCs (drepanocytes) due to the rod-shaped polymers of the inherited abnormal hemoglobin S (HbS). The RBC does not become sickled until it has lost its nucleus and has its full complement of HbS. Sickle cells are thin, elongated, and well filled with HbS. The main clinical manifestations of sickle cell disease are chronic hemolytic anemia and occlusion of microvasculature (called vaso-occlusive disease). Vaso-occlusive crisis may occur in the brain, liver, lung, or spleen. Factors that induce sickling are PO_2 (e.g., high altitude) or a concentration of 60% HbS or greater in RBCs. **(B) Thalassemia.** Photomicrograph shows microcytic and hypochromic RBCs due to various mutations that result in reduced amounts of hemoglobin A (HbA). Note the target cells (arrows). In addition, RBCs show anisocytosis (excessive variation in RBC size) and poikilocytosis (abnormal shapes of RBCs).

Review Test

1. A glutamic acid to valine substitution in the beta-globin subunit of hemoglobin causes which one of the following?

(A) α-thalassemia
(B) β-thalassemia
(C) sickle cell disease
(D) Hb Bart hydrops fetalis syndrome

2. Individuals with sickle cell disease appear healthy at birth. This is due to which one of the following?

(A) presence of HbF
(B) presence of HbS
(C) presence of HbA
(D) presence of HbD

3. Normal hematologic findings would be expected with which on of the following?

(A) α^0-thalassemia
(B) α^+-thalassemia
(C) thalassemia major
(D) β-thalassemia

4. A severe form of α-thalassemia that results in death in neonates is which one of the following?

(A) hemoglobin H disease
(B) Hb Bart
(C) thalassemia major
(D) thalassemia minor

5. Reduced or absent synthesis of β-globin subunits of hemoglobin results in which one of the following?

(A) β-thalassemia
(B) α^0-thalassemia
(C) α^+-thalassemia
(D) sickle cell anemia

Answers and Explanations

1. **The answer is (C).** Sickle cell disease results from a missense mutation in the second nucleotide of the sixth codon, inherited in an autosomal recessive fashion, which results in a glutamic acid to valine substitution in the β-globin subunit of hemoglobin. The resulting hemoglobin S (HbS) molecule forms polymers that distort the cells and leave them poorly deformable. The thalassemias, including Hb Bart result from deletions or mutations that cause reduced or absent synthesis of hemoglobin subunits.

2. **The answer is (A).** Fetal hemoglobin (Hb F) is the predominant form of hemoglobin in prenatal life and persists at high levels for some time after birth. Hb F does not contain β-globin subunits so it is unaffected by the mutation that causes sickle cell disease. Hb A is the normal adult hemoglobin. HbS and HbD are both abnormal hemoglobins caused by the common sickle cell mutation and coinheritance of another mutation in the same gene respectively.

3. **The answer is (B).** This is the silent carrier state for α-thalassemia, and in many cases, the hematologic findings are normal.

4. **The answer is (B).** Hb Bart is the most severe form of α-thalassemia and results from deletion or dysfunction of all four α-globin alleles.

5. **The answer is (A).** β-Thalassemia is defined by the absence or reduced synthesis of the β-globin subunits of hemoglobin.

chapter 14 | Genetics of Bleeding Disorders

I. HEMOPHILIA A (FACTOR VIII DEFICIENCY)

A. Hemophilia A is an **X-linked recessive** genetic disorder caused by a mutation in the *F8* gene on **chromosome Xq28** for **coagulation factor VIII.**

B. Factor VIII participates in the **intrinsic pathway of hemostasis** (blood clotting) in the following way: factor VIII binds to von Willebrand factor (vWF) in the circulation for stabilization (vWF acts as a carrier protein for factor VIII); factor VIII is cleaved by thrombin, released from vWF, and binds to activated platelet membranes where it interacts with factor IX; the factor VIII and factor IX interaction activates factor X (which is a critical step in early hemostasis).

C. Hemophilia A is most commonly (45% of cases) caused by the *F8* **intron 22-A gene inversion** (**"flip" inversion**) mutation. The remainder of cases are caused by missense, complete or partial deletions, RNA splicing, nonsense, large insertion, sequence duplications, and frameshift mutations.

D. Hemophilia A is defined by a reduced factor VIII clotting activity in the presence of normal vWF levels.

E. **Prevalence.** The prevalence of hemophilia A is 1/10,000 births in the US population.

F. There are three clinically significant forms of hemophilia A:
1. **Severe hemophilia A.**
   a. Severe hemophilia A results from <1% of factor VIII clotting activity.
   b. **Clinical features include:** usually diagnosed before 1 year of age, prolonged oozing after injuries, renewed bleeding after initial bleeding has stopped, delayed bleeding, large "goose eggs" after minor head bumps, abnormal bleeding after minor injuries, deep muscle hematomas, episodes of spontaneous joint bleeding are frequent, and 2 to 5 spontaneous bleeding episodes/month without adequate treatment.
2. **Moderately severe hemophilia A.**
   a. Moderately severe hemophilia A results from 1% to 5% of factor VIII clotting activity.
   b. **Clinical features include:** usually diagnosed before 5 to 6 years of age, prolonged oozing after injuries, renewed bleeding after initial bleeding has stopped, delayed bleeding, abnormal bleeding after minor injuries, episodes of spontaneous joint bleeding are rare, and one bleeding episode/month → one bleeding episode/year.
3. **Mild hemophilia A.**
   a. Mild hemophilia A results from 6% to 35% of factor VIII clotting activity.
   b. **Clinical features include:** usually diagnosed later in life, prolonged oozing after injuries, renewed bleeding after initial bleeding has stopped, delayed bleeding, abnormal bleeding after major injuries, episodes of spontaneous joint bleeding are absent, and 1 bleeding episode/year → 1 bleeding episode/10years.

II. HEMOPHILIA B (FACTOR IX DEFICIENCY; CHRISTMAS DISEASE)

A. Hemophilia B is an X-linked recessive genetic disorder caused by a mutation in the **F9** gene on **chromosome Xq27.1-q27.2** for **coagulation factor IX.**

B. Factor IX participates in the intrinsic pathway of hemostasis (blood clotting) in the following way: factor VIII is cleaved by thrombin, released from vWF, and binds to activated platelet membranes where it interacts with factor IX; the factor VIII and factor IX interaction activates factor X (which is a critical step in early hemostasis).

C. Hemophilia B is caused by a wide variety of mutations, which include: missense, complete or partial deletions, RNA splicing, nonsense, and frameshift mutations.

D. Hemophilia B is defined by a reduced factor IX clotting activity in the presence of normal vWF levels.

E. **Prevalence.** The prevalence of hemophilia B is 1/25,000 in the US population.

F. There are three clinically significant forms of hemophilia B:
  1. **Severe hemophilia B.**
      a. Severe hemophilia B results from <1% of factor IX clotting activity.
      b. **Clinical features include:** usually diagnosed before 1 year of age, prolonged oozing after injuries, renewed bleeding after initial bleeding has stopped, delayed bleeding, large "goose eggs" after minor head bumps, abnormal bleeding after minor injuries, deep muscle hematomas, episodes of spontaneous joint bleeding are frequent, and 2 to 5 spontaneous bleeding episodes/month without adequate treatment.
  2. **Moderately severe hemophilia B.**
      a. Moderately severe hemophilia B results from 1% to 5% of factor IX clotting activity.
      b. **Clinical features include:** usually diagnosed before 5 to 6 years of age, prolonged oozing after injuries, renewed bleeding after initial bleeding has stopped, delayed bleeding, abnormal bleeding after minor injuries, episodes of spontaneous joint bleeding are rare, and one bleeding episode/month → one bleeding episode/year.
  3. **Mild hemophilia B.**
      a. Mild hemophilia B results from 5% to 30% of factor IX clotting activity.
      b. **Clinical features include**: usually diagnosed later in life, prolonged oozing after injuries, renewed bleeding after initial bleeding has stopped, delayed bleeding, abnormal bleeding after major injuries, episodes of spontaneous joint bleeding are absent, and 1 bleeding episode/year → 1 bleeding episode/10 years.

III. VON WILLEBRAND DISEASE (VWD)

A. VWD is a genetic disorder caused by a mutation in the **VWF gene** on **chromosome 12p13.3** for **von Willebrand factor (vWF).**

B. VWD is the most common inherited bleeding disorder. However, only a small number of patients come to medical attention because of bleeding symptoms.

C. vWF participates in the intrinsic pathway of hemostasis (blood clotting) in the following way: vWf acts as a carrier protein for factor VIII; forms a bridge between vascular subendothelial connective tissue and platelets by binding to the **platelet receptor Gp1b** at sites of endothelial damage.

D. The majority of mutations causing VWD are still undefined. A database of mutations is available at http://*www.ragtimedesign.com/vwf/mutations*. A mutation that has been reported in a number of patients is a missense mutation, which results in a **normal cysteine → arginine** at position 386 (C386R).

E. VWD is defined by a reduced synthesis or reduced functionality of vWF.

F. Prevalence. The prevalence of VWD is 1/10,000 worldwide. The prevalence of VWD is 1/1,200 births in the Veneto region in northern Italy.

G. There are three clinically significant forms of VWD:
1. **Type 1 VWD.**
   a. Type 1 VWD is an autosomal dominant genetic disorder.
   b. Type 1 VWD results from a reduced synthesis of vW.
   c. Type 1 VWD accounts for 75% of the cases (i.e., the most common type of VWD).
   d. **Clinical features include:** can be diagnosed at any age; lifelong easy bruising; nose bleeding (epistaxis); skin bleeding; prolonged bleeding from mucosal surfaces; heavy menstrual bleeding; and mild to moderately severe bleeding symptoms, while some patients are asymptomatic.
2. **Type 2 VWD.**
   a. Type 2 VWD is an autosomal dominant or autosomal recessive genetic disorder.
   b. Type 2 VWD results from a reduced functionality of vWF.
   c. There are four subtypes of Type 2 VWD: 2A, 2B, 2M, and 2N.
   d. Type 2A VWD accounts for 10% to 15% of the cases.
   e. **Clinical features include:** can be diagnosed at any age, lifelong easy bruising, nose bleeding (epistaxis), skin bleeding, prolonged bleeding from mucosal surfaces, heavy menstrual bleeding, and moderate to moderately severe bleeding.
3. **Type 3 VWD.**
   a. Type 3 VWD is an autosomal recessive genetic disorder.
   b. Type 3 VWD results from a reduced synthesis of vWF and factor VIII.
   c. Type 3 VWD is a rare disease but the most severe form of VWD.
   d. Clinical features include: nose bleeding (epistaxis); severe skin bleeding; severe bleeding from mucosal surfaces; muscle hematomas; and severe joint bleeding.

IV. SUMMARY TABLE OF LABORATORY FINDINGS IN BLEEDING DISORDERS (Table 14-1)

t a b l e **14-1** Summary Table of Laboratory Findings in Bleeding Disorders

Lab Findings	Hemophilia A	Hemophilia B	von Willebrand Disease
Platelet count	Normal	Normal	Normal
Bleeding time	Normal	Normal	**Prolonged**
Prothrombin time	Normal	Normal	Normal
Partial thromboplastin time	Prolonged	Prolonged	Prolonged or normal
Factor VIII	**Low**	Normal	Normal
Factor IX	Normal	**Low**	Normal
v WF	Normal	Normal	**Low**
Ristocetin-induced platelet aggregation	Normal	Normal	**Impaired**

V. SUMMARY TABLE OF BLEEDING DISORDERS (Table 14-2)

table 14-2	Summary Table of Bleeding Disorders	
Genetic Disorder	**Gene/Gene Product Chromosome**	**Clinical Feature**
Hemophilia A	*F8* gene/factor VIII Xq28 F8 intron 22-A gene inversion ("flip" inversion) most common mutation Reduced factor VIII clotting activity	**Severe hemophilia A:** Usually diagnosed before 1 year of age; prolonged oozing after injuries; renewed bleeding after initial bleeding has stopped; delayed bleeding; large "goose eggs" after minor head bumps; abnormal bleeding after minor injuries; deep muscle hematomas; episodes of spontaneous joint bleeding are frequent; and 2–5 spontaneous bleeding episodes/month without adequate treatment. **Moderately severe hemophilia A:** Usually diagnosed before 5–6 years of age; prolonged oozing after injuries; renewed bleeding after initial bleeding has stopped; delayed bleeding; abnormal bleeding after minor injuries; episodes of spontaneous joint bleeding are rare; and 1 bleeding episode/month → 1 bleeding episode/year. **Mild hemophilia A:** Usually diagnosed later in life; prolonged oozing after injuries; renewed bleeding after initial bleeding has stopped; delayed bleeding; abnormal bleeding after major injuries, episodes of spontaneous joint bleeding are absent; and 1 bleeding episode/year → 1 bleeding episode/10 years.
Hemophilia B	*F9* gene/ factor IX Xq27.1-q27.2 Wide variety of mutations Reduced factor IX clotting activity	**Severe hemophilia B:** Usually diagnosed before 1 year of age; prolonged oozing after injuries; renewed bleeding after initial bleeding has stopped; delayed bleeding; large "goose eggs" after minor head bumps; abnormal bleeding after minor injuries; deep muscle hematomas; episodes of spontaneous joint bleeding are frequent; and 2 to 5 spontaneous bleeding episodes/month without adequate treatment. **Moderately severe hemophilia B:** Usually diagnosed before 5 to 6 years of age; prolonged oozing after injuries; renewed bleeding after initial bleeding has stopped; delayed bleeding; abnormal bleeding after minor injuries; episodes of spontaneous joint bleeding are rare; and 1 bleeding episode/month → 1 bleeding episode/year. **Mild hemophilia B:** Usually diagnosed later in life; prolonged oozing after injuries; renewed bleeding after initial bleeding has stopped; delayed bleeding; abnormal bleeding after major injuries; episodes of spontaneous joint bleeding are absent; and 1 bleeding episode/year → 1 bleeding episode/10 years.
von Willebrand Disease	*VWF* gene/ von Willebrand factor 12p13.3 Wide variety of mutations Reduced synthesis or functionality of vWF	**Type 1 VWD:** Can be diagnosed at any age; lifelong easy bruising; nose bleeding (epistaxis); skin bleeding; prolonged bleeding from mucosal surfaces; heavy menstrual bleeding; and mild to moderately severe bleeding symptoms while some patients are asymptomatic. **Type 2 VWD:** can be diagnosed at any age; lifelong easy bruising; nose bleeding (epistaxis); skin bleeding; prolonged bleeding from mucosal surfaces; heavy menstrual bleeding; and moderate to moderately severe bleeding. **Type 3 VWD:** Nose bleeding (epistaxis); severe skin bleeding; severe bleeding from mucosal surfaces; muscle hematomas; and severe joint bleeding.

Review Test

1. Reduced factor VIII clotting activity with normal von Willebrand factor is a finding in which one of the following?

(A) von Willebrand disease
(B) hemophilia B
(C) Christmas disease
(D) hemophilia A

2. A gene inversion, called a "flip" inversion, is the most common mutation in which one of the following?

(A) hemophilia A
(B) hemophilia B
(C) von Willebrand disease
(D) Christmas disease

3. Which one of the following inherited bleeding disorders is autosomal dominant?

(A) hemophilia A
(B) hemophilia B
(C) von Willebrand disease
(D) Christmas disease

4. The bleeding disorder that is most likely to be asymptomatic is which one of the following?

(A) hemophilia A
(B) hemophilia B
(C) Type 1 von Willebrand disease
(D) Type 3 von Willebrand disease

5. The most common bleeding disorder is which one of the following?

(A) hemophilia A
(B) hemophilia B
(C) von Willebrand disease
(D) Christmas disease

Answers and Explanations

1. **The answer is (D).** Hemophilia A is caused by a mutation in the F8 gene for coagulation factor VIII.

2. **The answer is (A).** An F8 intron gene inversion, called a "flip" inversion, is responsible for the majority cases of hemophilia A.

3. **The answer is (C).** Von Willebrand disease is the most common inherited bleeding disorder.

4. **The answer is (C).** The bleeding disorder most likely to be asymptomatic is Type 1 von Willebrand disease because there is von Willebrand factor present but in reduced amounts.

5. **The answer is (C).** The most common inherited bleeding disorder is von Willebrand disease, but only a small number of patients come to medical attention because of symptoms.

I. CAUSES OF HUMAN BIRTH DEFECTS (Table 15-1)

t a b l e **15-1** Causes of Human Birth Defects	
Causes	**Percentage**
Unknown factors	45%
Multifactorial inheritance (environmental and genetic causes combined)	25%
Chromosome abnormalities (numerical or structural)	10%
Mendelian single gene inheritance	5%
Teratogen exposure	5%
Uterine factors (e.g., oligohydramnios, uterine fibroids)	3%
Twinning	1%

II. TYPES OF HUMAN BIRTH DEFECTS

A. **Malformation (genetic based).** A morphological defect caused by an **intrinsically abnormal developmental process**. Intrinsic implies that the developmental potential of the primordium is abnormal from the beginning (e.g., a chromosome abnormality of a gamete at fertilization). A malformation occurs during the embryonic period (weeks 3 to 8 of gestation) when all major organ systems begin to develop (i.e., organogenesis). Malformation may also be due to nutritional deficiencies (e.g., lack of folate in neural tube defects). Malformation examples include: polydactyly, oligodactyly, spina bifida, cleft palate, and most kinds of congenital heart malformations.

B. **Dysplasia (genetic based).** A morphological defect caused by an **abnormal organization of cells into tissues.** A dysplasia occurs during the embryonic period (weeks 3 to 8 of gestation) when all major organ systems begin to develop (i.e., organogenesis). Dysplasia examples include: thanatophoric dwarfism and congenital ectodermal dysplasia.

C. **Disruption (not genetic based).** A morphological defect caused by the breakdown of or interference with an intrinsically normal developmental process. A disruption occurs at any time during gestation. Disruption examples include: bowel atresia due to vascular accidents, amniotic band disruptions, and most cases of porencephaly (cystic lesions of the brain).

D. **Deformation (not genetic based).** An abnormality of form or position of a body part caused by mechanical forces that interfere with normal growth or position of the fetus in utero. A deformation occurs usually in the second and third trimester of the pregnancy. Deformation

153

examples include: abnormal position of the feet, clubfoot, and abnormal moulding of the head.

III. PATTERNS OF HUMAN BIRTH DEFECTS

When a patient presents with multiple birth defects, the following patterns may be presented:

A. Sequence. A pattern of multiple defects derived from a single known or presumed structural defect or mechanical factor. Sequence examples include Robin sequence and oligohydramnios sequence.

B. Syndrome. A pattern of multiple defects all of which are pathogenetically related. In clinical genetics, "syndrome" implies a similar cause in all affected individuals. Syndrome examples include Down syndrome, fetal alcohol syndrome, and Marfan syndrome.

C. Polytopic Field Defect. A pattern of multiple defects derived from the disturbance of a single developmental field. Developmental fields are regions of the embryo that develop in a related fashion although the derivative structures may not be close spatially in the infant or adult. Polytopic field defect examples include holoprosencephaly.

D. Association. A pattern of multiple defects that occurs more often than expected by chance alone (i.e., nonrandom) but has not yet been classified as a sequence, syndrome, or polytopic field defect. As development genetics advances, many associations will very likely be reclassified as a sequence, syndrome, or polytopic defect. Association examples include abnormal ears associated with renal defects; single umbilical artery associated with heart defects; and, the association of vertebral, heart, and kidney defects.

IV. DETERMINATION OF THE LEFT/RIGHT (L/H) AXIS

L/R axis determination is established early in embryological development and is caused by a cascade of paracrine signaling proteins.

A. L/R axis determination begins with the asymmetric (future left-side) expression of the signaling protein **Sonic hedgehog (Shh) protein** from the notochord, which is located in the midline.

B. This results in the expression of the signaling protein **nodal protein** (a member of the TGF-β family) only on the left side of the embryo (left side = nodal positive; right side = nodal negative) and may be the earliest event in L/R axis determination.

C. After the L/R axis is determined in the embryo, the L/R asymmetry of a number of anatomical organs (e.g., heart, liver, stomach) can then be patterned under the influence of the transcription factor **zinc-finger protein of the cerebellum (ZIC3)**.

D. Clinical consideration: **Primary Ciliary Dyskinesia (PCD; or Immotile Cilia Syndrome)/ Kartagener Syndrome.**

1. PCD is an autosomal recessive genetic disorder caused by missense, nonsense, splice site, insertion, and deletion mutations where at least two different genes have been implicated thus far:

   a. ***DNAH5* gene** on **chromosome 5p15-p14** for **ciliary *dy*nein *a*xonemal *h*eavy chain 5**. This mutation occurs in 28% of the cases.

   b. ***DNAI1* gene** on **chromosome 9p21-p13** for ***dy*nein *a*xonemal *i*ntermediate chain 1**. This mutation occurs in 10% of the cases.

   c. ≈60% of PCD affected individuals do not have mutations in the *DNAH5* gene or *DNAI1* gene. It is speculated that mutations in other genes on chromosomes 15q24-25, 15q13.1-q15.1, 16p12.1-p12.2, and 19q13.42-q13.43 for dynein light chains, spoke head proteins, and other axonemal proteins may be causative.

2. These mutations result in defective outer dynein arms that results in cilia that are immotile (ciliary immotility), beat abnormally (ciliary dyskinesia), or are absent (ciliary aplasia).

3. PCD affected individuals inherit the mutant genes from the parents who are obligate asymptomatic heterozygotes.

4. **Prevalence.** The prevalence of PCD is 1/12,000 to 17,000 births in the US population.

5. **Clinical features include**: chronic cough; chronic rhinitis; chronic sinusitis; chronic/recurrent ear infections; recurrent sinus/pulmonary infections due to a defect of cilia in the respiratory pathways; neonatal respiratory distress; digital clubbing; sterility in males (retarded sperm movement); situs inversus totalis (mirror-image reversal of all visceral organs with no apparent consequences; PCD with situs inversus totalis is called **Kartagener syndrome**); heterotaxy (discordance of right and left patterns of ordinarily asymmetrical structures with significant malformations; for example asplenia or polysplenia); the gold standard diagnostic test is the appearance of ciliary ultrastructural defects obtained by electron microscopy of a respiratory epithelium biopsy.

V. DETERMINATION OF THE ANTERIOR/POSTERIOR (A/P) AXIS

A/P axis determination is established by the formation of the **primitive streak,** which involves the expression of the signaling protein **nodal protein** (a member of the TGF-β family).

A. A large number of gene regulatory proteins called **homeodomain proteins** play a role in determining the normal A/P location of a number of anatomical structures.

B. All homeotic genes encode for homeodomain proteins, which are gene regulatory proteins. Homeotic genes contain a 180 base pair sequence (called a **homeobox**) that encodes a 60 amino acid long region (called a **homeodomain**) that binds specifically to DNA segments.

C. A **homeotic mutation** is one in which one body part is substituted for another. Homeotic mutations were first studied in *Drosophila* (e.g., legs sprout from the head in place of antennae). The genes involved in homeotic mutations are called **homeotic genes**, which are collectively referred to as the ***HOM*-complex.**

D. **Clustered Human Homeotic Genes.** There are 39 clustered homeotic genes identified in humans thus far. They are organized into four gene clusters (***HoxA, HoxB, HoxC, and HoxD***) collectively called the ***Hox*-complex**. In addition, there are numerous **nonclustered homeotic genes** randomly dispersed throughout the human genome.

VI. GROWTH AND DIFFERENTIATION (Figure 15-1)

The close range interaction between two or more cells or tissues of different histories is called **induction**. Induction involves an **inducer** (a cell or tissue that produces a signal that changes the behavior of another cell or tissue) and a **responder** (a cell or tissue that is induced). The inducer and responder may interact by either **juxtacrine interactions** or **paracrine interactions**. Juxtacrine interactions occur when cell membrane receptors on the inducer interact with cell membrane receptors on the responder. Paracrine interactions occur when the inducer secretes a protein that diffuses across a small distance and binds to a cell membrane receptor on the responder. These diffusible proteins are called **paracrine factors** or **growth and differentiation factors (GDFs).** When a paracrine factor binds to a cell membrane receptor on the responder a series of reactions occurs called a **signal transduction pathway**. The end point of a signal transduction pathway is either **activation or deactivation of transcription factors** (i.e., the responder expresses different genes) or the **regulation of the cytoskeleton** (responder changes shape or is permitted to migrate). An important family of paracrine factors is the fibroblast growth factor (FGF) family as indicated below.

A. **Fibroblast Growth Factor (FGF) Family.** The FGF family has nine members **(FGF1-FGF9)** along with a number of isoforms. FGFs bind to **FGF receptors (FGFRs).** The FGFR family has four members **(FGFR1-FGFR4).** FGFRs are highly homologous glycoproteins with a signal peptide domain, three immunoglobulinlike domains (IgI-IgIII), an acid box domain, a transmembrane domain, and two intracellular tyrosine kinase domains (i.e., **receptor tyrosine kinases**). FGFRs subsequently act through two major signal transduction pathways called the **receptor tyrosine kinase pathway (RTK pathway)** and the *Janus kinase –signal transducers and activators of transcription pathway (JAK/STAT pathway).*

B. **Clinical Considerations.**
1. **Achondroplasia (AC). Skeletal dysplasias** are conditions of abnormal bone growth and are typically called **dwarfisms**. There are **short-limb dysplasias** (short limbs relative to the length of the trunk) and **short-trunk dysplasias** (short trunk relative to the length of the limbs). AC is a short-limb dysplasia.
   a. AC is an autosomal dominant genetic disorder caused by a missense mutation in the *FGFR3* gene on **chromosome 4p16.3** for the **fibroblast growth factor receptor 3**.
   b. The most common mutation is a G → A transition at **nucleotide position 1138 (G1138A)** which results in a **normal glycine → arginine** substitution at position 380 (G380R) in the **transmembrane domain** of FGFR3.
   c. This mutation results in **constitutive activation** of FGFR3 (i.e., a **gain-of-function mutation**) which indicates that FGFR3 normally inhibits bone growth.
   d. Most AC affected individuals have a *de novo* mutation. The *de novo* mutation usually occurs during spermatogenesis in the unaffected advanced-aged father. Chances of AC increase with increasing paternal age.
   e. **Prevalence.** The prevalence of AC is 1/26,000 to 40,000 births. This is the most common type of dwarfism.
   f. **Clinical features include:** short stature, proximal shortening of arms and legs with redundant skin folds, limitation of elbow extension, trident configuration of hands, bow legs, thoracolumbar gibbus in infancy, exaggerated lumbar lordosis, large head with frontal bossing, and midface hypoplasia; mental function is not affected.
2. **Hypochondroplasia (HP).** HP is a short-limb dysplasia.
   a. HP is an autosomal dominant genetic disorder caused by a missense mutation in the *FGFR3* gene on **chromosome 4p16.3** for the **fibroblast growth factor receptor 3**.
   b. One of the most common mutations is a C → A transition at **nucleotide position 1620 (C1620A)** which results in a **normal asparagine → lysine** substitution at position 540 (N540K) in the first **tyrosine kinase domain** of FGFR3.

    c. This mutation results in **constitutive activation** of FGFR3 (i.e., a **gain-of-function muta-tion**) which indicates that FGFR3 normally inhibits bone growth.

    d. Most HP affected individuals have a *de novo* mutation. The de novo mutation usually occurs during spermatogenesis in the unaffected advanced-age father. Chances of HP increase with increasing paternal age.

    e. Prevalence. The prevalence of HP is 1/15,000 to 40,000 births.

    f. Clinical features include: short stature, stocky build, disproportionately short arms and legs, broad, short hands and feet, mild joint laxity, and macrocephaly. **The skeletal features are very similar to AC but generally more mild with less craniofacial involvement.**

3. Thanatophoric dysplasia (TD). TD is a short-limb dysplasia and is the **most common of the "lethal skeletal dysplasias."** There are two types of TD.

    a. Type 1 TD is an autosomal dominant genetic disorder caused by a missense mutation in the ***FGFR3* gene** on **chromosome 4p16.3** for the ***fibroblast growth factor receptor 3*.** This results in a **normal arginine → cysteine** substitution at position 248 (R248C) in the **region between IgII-IgIII domains** of FGFR3.

    b. Type 2 TD is an autosomal dominant genetic disorder caused by a mutation in the ***FGFR3* gene** on **chromosome 4p16.3** for the ***fibroblast growth factor receptor 3*.** This results in a **normal lysine → glutamic acid** substitution at position 650 (K650E) in the second **tyrosine kinase domain** of FGFR3.

    c. These mutations result in **constitutive activation** of FGFR3 (i.e., a **gain-of-function muta-tion**) which indicates that FGFR3 normally inhibits bone growth.

    d. Most TD affected individuals have a *de novo* mutation. The *de novo* mutation usually occurs during spermatogenesis in the unaffected advanced-age father. Chances of TD increase with increasing paternal age.

    e. Prevalence. The prevalence of TD is 1/20,000 to 50,000 births.

    f. Clinical features include: short ribs, narrow thorax, macrocephaly, distinctive facial fea-tures, brachydactyly, hypotonia, and redundant skin folds along the limbs; **children with TD usually die in the perinatal period** with only a few survivors into early childhood; a narrow thoracic cage which leads to respiratory compromise; curved long bones; type 1 TD is characterized by micromelia with bowed femurs and generally without a clover-leaf-shaped skull; type 2 TD is characterized by micromelia with straight long bones and generally with a cloverleaf-shaped skull.

4. Crouzon syndrome (CR). CR is one of eight FGFR-related craniosynostosis syndromes, which include Pfeiffer syndrome, Apert syndrome, Beare-Stevenson syndrome, FGFR2-related isolated coronal synostosis, Jackson-Weiss syndrome, Crouzon syndrome with acanthosis nigricans, and Muenke syndrome.

    a. CR is an autosomal dominant genetic disorder caused by a missense mutations in the ***FGFR2* gene** on **chromosome 10q25-q26** for **fibroblast growth factor receptor 2**.

    b. These missense mutations include a **normal cysteine → tyrosine** substitution at position 342 (C342Y), a **normal cysteine → arginine** substitution at position 342 (C342R), a **normal cysteine → tryptophan** substitution at position 342 (C342W), or a **normal cysteine → phenylalanine** substitution at position 278 (C278F) in the **IgIII domain** of FGFR2 (the so-called "cysteine mutational hotspot").

    c. These missense mutations result in **constitutive activation** of FGFR2 (i.e., a **gain-of-function mutation**) which indicates that FGFR2 normally inhibits bone growth.

    d. Most CR affected individuals inherit a mutant gene from an affected parent whereas some CR affected individuals have a *de novo* mutation. The *de novo* mutation usually occurs during spermatogenesis in the unaffected advanced-aged father. Chances of CR increase with increasing paternal age.

    e. Prevalence. The prevalence of CR is 1/62,500 births. The prevalence for all forms of FGFR-related craniosynostosis syndromes 1/2,000 to 2,500 births.

    f. Clinical features of CR include: premature craniosynostosis, midface hypoplasia with shallow orbits, ocular proptosis, mandibular prognathism, normal extremities, pro-gressive hydrocephalus, and no mental retardation.

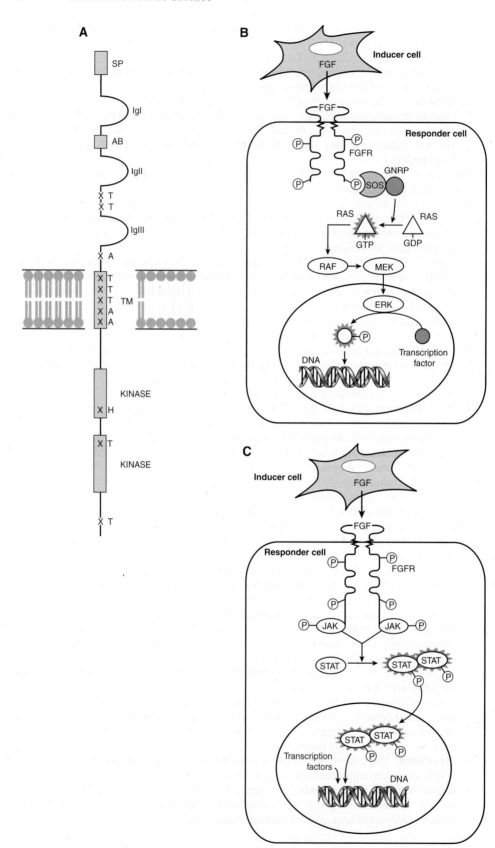

VII. FORMATION OF THE EXTRACELLULAR MATRIX (ECM)

The ECM consists of various macromolecules secreted by cells (e.g., mesenchymal cells, fibroblasts, chondroblasts, osteoblasts) and forms a noncellular material in the interstices between cells. The ECM is not inert but instead plays an important embryological role in cell adhesion, cell migration, and formation of epithelial sheets. The ECM consists of proteoglycans, glycoproteins, and fibers (i.e., collagen and elastic fibers).

A. Proteoglycans. Proteoglycans bind paracrine factors (e.g., FGF, Shh, Wnt, and TGF-β superfamily) secreted by an inducer cell and deliver the paracrine factors in high concentration to their respective receptors located on the responder cell. Specific proteoglycans include the following: **aggrecan, betaglycan, decorin, perlecan**, and **syndecan-1**.

B. Glycoproteins. Glycoproteins play a role in cell migration and modulation of gene expression activity. Specific glycoproteins include the following: **fibronectin, laminin, chondronectin, osteocalcin, osteopontin**, and **bone sialoprotein**.

C. Collagen and Elastic Fibers. Collagen is a family of proteins consisting of three polypeptide **α-chains** that form a triple-stranded helical structure. There are 25 distinct collagen α-chain genes. However, only ≈20 different types of collagens (types I-XX) have been isolated. **Elastic fibers** consist of an amorphous core of the **elastin** protein surrounded by microfibrils of the **fibrillin** protein.

1. **Osteogenesis imperfecta (OI).** OI is a group of disorders (types I-VII) with a continuum ranging from perinatal lethality → severe skeletal deformities → nearly asymptomatic individuals.

   a. OI (types I-IV) are autosomal dominant genetic disorders caused by mutations where at least two different genes have been implicated thus far:

   i. *COL1A1* gene on **chromosome 17q21.3-q22** for *co*llagen proα-1 (*I*) chain of Type I procollagen

   ii. *COL1A2* gene on **chromosome 7q22.1** for *co*llagen proα-2(*I*) chain of Type I procollagen.

   b. Type I OI is most commonly caused by a **frameshift mutation** or a **RNA splicing mutation** (that forms a premature STOP codon, shifts a reading frame, or produces unstable mRNAs).

   c. Types II, III, and IV OI are most commonly caused by a **missense mutation** which results in a normal glycine → serine, normal glycine → arginine, normal glycine → cysteine, or normal glycine → tryptophan substitution which alters the structure of the α1(I)-chain or the α2(I)-chain since glycine is necessary for normal folding of the collagen helix.

FIGURE 15-1. Fibroblast growth factor (FGF) receptor 3 and signal transduction pathways. (A) Diagram of the FGF receptor 3. Diagram shows important functional domains, which include a signal peptide (SP), three immunoglobulinlike domains (Ig), and acid box (AB), a transmembrane domain (TM), and two tyrosine kinase domains (Kinase). The location of various point mutations (x's) causing achondroplasia (A), hypochondroplasia (H), and thanatophoric dysplasia (T) are indicated. **(B) RTK (receptor tyrosine kinase) pathway.** When FGF (fibroblast growth factor) binds to the FGFR (fibroblast growth factor receptor), autophosphorylation of FGFR occurs. This is recognized by SOS adaptor protein, which activates GNRP (guanine nucleotide releasing factor). GNRP (guanine nucleotide releasing factor) activates the G protein RAS by exchanging a PO_4^{2-} from GTP to transform the bound GDP to GTP (RAS-GDP → RAS-GTP). RAS-GTP activates the kinase RAF, which activates the kinase MEK, which then activates the kinase ERK. ERK enters the nucleus and phosphorylates transcription factors, which then modulate gene expression activity. The flow of RTK pathway is: FGF → RTK → GNRP → RAS → RAF → MEK → ERK → Transcription factors. **(C) JAK (Janus kinase)/ STAT (signal transducers and activators of transcription) pathway.** When FGF binds to other receptors linked to JAK, the receptors dimerize and the JAK proteins phosphorylate each other and the dimerized receptors, which activates the dormant kinase activity of the receptor. The activated receptor phosphorylates STAT, which allows STAT to dimerize. The activated dimerized STAT enters the nucleus and along with other transcription factors modulates gene expression activity. The flow of the JAK/STAT pathway is: FGF → FGFR → JAK → STAT → STAT dimerization → Transcription factors.

d. ≈40% of Type I OI affected individuals (i.e., milder forms of OI) inherit a mutant gene from an affected parent whereas ≈60% have a *de novo* mutation. ≈100% of type II and III OI affected individuals (i.e., more severe forms of OI) have a *de novo* mutation.

e. Prevalence. The prevalence of all types of OI is 1/15,000 births in a Finnish population.

f. Clinical features include: extreme bone fragility with spontaneous fractures, short stature with bone deformities, grey or brown teeth, **blue sclera of the eye**, and progressive postpubertal hearing loss. Milder forms of OI may be confused with child abuse. Severe forms of OI are fatal *in utero* or during the early neonatal period.

2. Classic-type Ehlers Danlos syndrome (EDS).

a. EDS is an autosomal dominant genetic disorder caused by mutations where at least two different genes have been implicated thus far:

i. *COL5A1* gene on chromosome 9q34.2-q34.3 for *co*llagen proα-*1* (*V*) chain of Type V procollagen

ii. *COL5A2* gene on chromosome 2q31 for *co*llagen proα-*2* (*V*) chain of Type V procollagen

b. EDS is most commonly caused by a **nonsense mutation, frameshift mutation**, or a **RNA splicing mutation** (that forms a premature STOP codon, shifts a reading frame, or produces unstable mRNAs).

c. ≈50% of EDS affected individuals inherit a mutant gene from an affected parent whereas ≈50% have a *de novo* mutation.

d. Prevalence. The prevalence of EDS is 1/20,000 births.

e. Clinical features include: extremely stretchable and fragile skin, hypermobile joints, aneurysms of blood vessels, rupture of the bowel, abnormal wound healing, and widened atrophic scars.

3. Marfan syndrome (MFS).

a. MFS is an autosomal dominant genetic disorder caused by a mutation in the *FBN1* gene on **chromosome 15q21.1** for the *fi*brilli*n*-1 protein, which is an essential component of **microfibrils** found in both elastic and nonelastic tissue. Microfibrils play a role in the formation of the elastic matrix (i.e., **elastic fibers**), elastic matrix-cell attachments, and the regulation of growth factors.

b. MFS is caused by >200 different mutations and no common mutation is associated with any population.

c. ≈75% of MFS affected individuals inherit a mutant gene from an affected parent whereas ≈25% have a *de novo* mutation.

d. Prevalence. The prevalence of MFS is 1/5,000 to 10,000 births.

e. Clinical features include: unusually tall individuals, exceptionally long, thin limbs, pectus excavatum ("hollow chest"), scoliosis, **ectopia lentis (dislocation of the lens),** severe near-sightedness (myopia), **dilatation or dissection of the aorta** at the level of the sinuses of Valsalva which may lead to cardiomyopathy or even a rupture of the aorta, dural ectasia, and mitral valve prolapse.

VIII. NEURAL CREST CELL MIGRATION

The neural crest cells differentiate from cells located along the lateral border of the neural plate, which is mediated by **BMP-4** and **BMP-7 (body morphogenetic protein).** The differentiation of neural crest cells is marked by the expression of *slug* (a zinc-finger transcription factor) which characterizes cells that break away from the neuroepithelium of the neural plate and migrate into the extracellular matrix as mesenchymal cells. Neural crest cells undergo a prolific migration throughout the embryo (both the cranial region and trunk region) and ultimately differentiate into a wide array of adult cells and structures. **Neurocristopathy** is a termed used to describe any disease related to maldevelopment of neural crest cells.

A. Waardenburg syndrome Type 1 (WS Type 1).

1. WS type 1 is an autosomal dominant genetic disorder caused by mutation in the **PAX3 gene** on **chromosome 2q35** for the **paired box protein PAX3**. The *PAX* genes are characterized by a 128 amino acid DNA-binding domain called a paired box.

2. Paired box protein PAX3 is one of a family of nine human PAX genes coding for **DNA-binding transcription factors** that are expressed in the early embryo and regulate neural crest-derived cell types, including melanocytes. .

3. The mutations of the *PAX3* gene include **missense, nonsense, frameshift, whole gene deletions, intragenic deletions,** and **RNA splicing mutations**; all of which result in a **loss-of-function mutation**.

4. ≈90% of WS affected individuals inherit a mutant gene from an affected parent whereas ≈10% have a *de novo* mutation. The *de novo* mutations usually occur during spermatogenesis in the unaffected advanced-aged father. Chances of WS type 1 increase with increasing paternal age.

5. **Prevalence.** The prevalence of WS type 1 is 1/20,000 to 40,000 births. WS type 1 is responsible for ≈3% of congenitally deaf children.

6. **Clinical features include:** dystopia canthorum (lateral displacement of the inner canthi), growing together of eyebrows, lateral displacement of lacrimal puncta, a broad nasal root, heterochromia of the iris, congenital deafness or hearing impairment, and piebaldism including a white forelock and a triangular area of hypopigmentation.

B. Nonsyndromic Congenital Intestinal Aganglionosis (Hirschsprung disease; HSCR).

1. Nonsyndromic HSCR is an autosomal dominant genetic disorder associated with mutations where at least six different genes have been implicated thus far:

    a. **RET (rearranged during transfection) gene** on **chromosome 10q11.2** for a **receptor tyrosine kinase** (≈90% of HSCR cases)

    b. **GDNF gene** on **chromosome 5p13.1-p14** for **glial cell line-derived neurotrophic factor**

    c. **NRTN gene** on **chromosome 19p13.3** for **neurturin**

    d. **EDNRB gene** on **chromosome 13q22** for the **endothelin B receptor**

    e. **EDN3 gene** on **chromosome 20q13.2-q13.3** for **endothelin-3**

    f. **ECE1** on **chromosome 1p36.1** for **endothelin-converting enzyme**

2. RET protein is expressed by enteric neuronal precursor cells after they leave the neural tube and throughout their colonization of the gut tube. RET ligands are **GDNF** and **NRTN** which are expressed by nearby mesenchymal cells.

3. The mutations in the *RET* gene result in a **loss-of-function mutation**.

4. A significant proportion of HSCR affected individuals inherit a mutant gene from a completely *unaffected* parent (because of incomplete penetrance or variable expressivity the parent is unaffected resulting in what appears to be a negative family history). A small proportion of HSCR affected individuals inherit two different mutant genes with each parent contributing a single mutant allele (i.e., digenic inheritance). The proportion of HSCR affected individuals with a *de novo* mutation is unknown.

5. **Prevalence.** The prevalence of HSCR is 1/5000 births. 80% of HSCR affected individuals have aganglionosis restricted to the rectosigmoid colon ("short segment disease"). Fifteen percent to 20% of HSCR affected individuals have aganglionosis that extends proximal to the sigmoid colon ("long segment disease"). Two percent to 10% of HSCR affected individuals have trisomy 21 (Down syndrome). This association between trisomy 21 and HSCR remains unexplained.

6. **Clinical findings include:** arrest of the caudal migration of neural crest cells resulting in the absence of ganglionic cells in the myenteric and submucosal plexuses; abdominal pain and distension; **inability to pass meconium within the first 48 hours of life;** gushing of fecal material upon a rectal digital exam; constipation; emesis; a loss of peristalsis in the colon segment distal to the normal innervated colon; and the failure of internal anal sphincter to relax following rectal distention (i.e., abnormal rectoanal reflex).

C. Orofacial clefting.

1. Orofacial clefting is a multifactorial genetic disorder associated with mutations where at least five different genes have been implicated thus far:

a. **DLX 1-6** (**d**istal **l**ess homeobox) **gene family** on **chromosome 2q32, 2cen-q33,17q21.3-q22,17q21.33,7q22,** and **7q22,** respectively, for various **homeodomain gene regulatory proteins**

b. **SHH** (**s**onic **h**edge**h**og) **gene** on **chromosome 7q36** for the **shh protein**

c. **TGF-α** (**t**ransforming **g**rowth **f**actor) **gene** on **chromosome 2p13** for the **TGF-α variant protein**

d. **TGF-β gene** on **chromosome 14q24** for the **TGF-β protein**

e. **IRF-6** (**i**nterferon **r**egulatory **f**actor) **gene** on **chromosome 1q32.3-q41** for the **IRF6 transcription factor.**

2. Orofacial clefting has a **multifactorial inheritance** (i.e., a genetic component interacting with the environment). Orofacial clefting is a genetically complex event; a single gene mutation causing orofacial clefting probably does not occur.

3. Environmental factors that may play a role in orofacial clefting involve exposure of the fetus to **phenytoin, sodium valproate, methotrexate,** and **folate deficiency**.

4. The most common craniofacial birth defect is the orofacial cleft, which consists of **cleft lip with or without cleft palate (CL/P)** or **isolated cleft palate (CP).** CL/P and CP are distinct birth defects based on their embryological formation, etiology, candidate genes, and recurrence risk.

5. **Prevalence.** CL/P is more common than CP and varies by ethnicity. The prevalence of CL/P is 1/500 births in American Indian and Asian populations, 1/1,000 births in Caucasian populations, and 1/2,000 in the African American population. CL/P occurs more frequently in males. The prevalence of CP is 1/2,500 births occurs in 2,500 births, does not show ethnic variation, and occurs more frequently in females.

6. **Clinical features include:** Cleft lip (unilateral cleft lip on the left side is most commonly seen); cleft palate; a combination of both cleft lip and cleft palate; and other associated abnormalities in the central nervous system; skeletal system; cardiovascular system; and all tissue of neuroectodermal origin.

D. Treacher Collins Syndrome (or Mandibulofacial Dysostosis; TCS). TCS belongs to a category of **first arch syndromes,** which result from a lack of neural crest cell migration into pharyngeal arch 1, and produces various facial anomalies. There are two well-described first arch syndromes: TCS and **Pierre Robin syndrome**.

1. TCS is an autosomal dominant genetic disorder caused by a mutation in the **TCOFI gene** on **chromosome 5q32-q33.1** for the **treacle protein.**

2. The treacle protein is a nucleolar protein related to the nucleolar phosphoprotein **Nopp140** both of which contain **LIS1 motifs** leading to the speculation of **microtubule dynamics** involvement. In addition, treacle interacts with the small nucleolar ribonucleoprotein **hNop56p** leading to the speculation of **ribosomal biogenesis** involvement.

3. >100 mutations in the *TCOF1* gene have been identified with **frameshift mutations** forming a premature STOP codon, which is the most common type of mutation. The mutations in the *TCOFI* gene result in a **loss-of-function mutation**.

4. ≈40% of TCS affected individuals inherit a mutant gene from an affected parent whereas ≈60% have a *de novo* mutation.

5. **Prevalence.** The prevalence of TCS is 1/10,000-50,000 births.

6. **Clinical features include:** hypoplasia of the zygomatic bones and mandible resulting in midface hypoplasia, micrognathia, and retrognathia; external ear abnormalities including small, absent, malformed, or rotated ears; and lower eyelid abnormalities including coloboma.

IX. SUMMARY TABLE OF DEVELOPMENTAL DISORDERS (Table 15-2)

Genetic Disorder	Gene/Gene Product Chromosome	Clinical Features
Determination of left/right axis		
Primary ciliary dyskinesia	*DNAH5* gene/ciliary dynein axonemal heavy chain 5 5p15-p14 *DNAI1* gene/dynein axonemal intermediate chain 1 9p21-p13	Chronic cough; chronic rhinitis; chronic sinusitis; chronic/recurrent ear infections; recurrent sinus/pulmonary infections due to a defect of cilia in the respiratory pathways; neonatal respiratory distress; digital clubbing; sterility in males (retarded sperm movement); situs inversus totalis (mirror-image reversal of all visceral organs with no apparent consequences; PCD with situs inversus totalis is called **Kartagener syndrome**); heterotaxy (discordance of right and left patterns of ordinarily asymmetrical structures with significant malformations; for example asplenia or polysplenia); the gold standard diagnostic test is the appearance of ciliary ultrastructural defects obtained by electron microscopy of a respiratory epithelium biopsy.
Growth and differentiation		
Achondroplasia	*FGFR3* gene/fibroblast growth factor receptor 3 4p16.3	Short stature, proximal shortening of arms and legs with redundant skin folds, limitation of elbow extension, trident configuration of hands, bow legs, thoracolumbar gibbus in infancy, exaggerated lumbar lordosis, large head with frontal bossing, and midface hypoplasia, mental function is not affected.
Hypochondroplasia	*FGRR3* gene/fibroblast growth factor receptor 3 4p16.3	Short stature, stocky build, disproportionately short arms and legs, broad, short hands and feet, mild joint laxity, and macrocephaly. The skeletal features are very similar to AC but generally are more mild with less craniofacial involvement.
Thanatophoric dysplasia	*FGFR3* gene/fibroblast growth factor receptor 3 4p16.3	Short ribs, narrow thorax, macrocephaly, distinctive facial features, brachydactyly, hypotonia, and redundant skin folds along the limbs; children with TD usually die in the perinatal period with only a few survivors into early childhood; a narrow thoracic cage which leads to respiratory compromise; curved long bones; type 1 TD is characterized by micromelia with bowed femurs and generally without a cloverleaf-shaped skull; type 2 TD is characterized by micromelia with straight long bones and generally with a cloverleaf-shaped skull.
Crouzon syndrome	*FGFR2* gene/fibroblast growth factor receptor 2 10q25-q26	Premature craniosynostosis, midface hypoplasia with shallow orbits, ocular proptosis, mandibular prognathism, normal extremities, progressive hydrocephalus, and no mental retardation.

(continued)

table	**15-2**	(continued)	

Genetic Disorder	Gene/Gene Product Chromosome	Clinical Features
Formation of the extracellular matrix		
Osteogenesis imperfecta	*COL1A1* gene/collagen proα-1(I) chain of Type I procollagen 17q21.3-q22 *COL1A2* gene/collagen proα-2 (I) chain of Type I procollagen 7q22.1	Extreme bone fragility with spontaneous fractures, short stature with bone deformities, grey or brown teeth, blue sclera of the eye, and progressive postpubertal hearing loss. Milder forms of OI may be confused with child abuse. Severe forms of OI are fatal *in utero* or during the early neonatal period.
Classic-type Ehlers Danlos syndrome	*COL5A1* gene/collagen proα-1 (V) chain of Type 5 procollagen 9q34.2-q34.3 *COL5A2* gene/collagen proα-2(V) chain of Type V procollagen 2q31	Extremely stretchable and fragile skin, hypermobile joints, aneurysms of blood vessels, rupture of the bowel, abnormal wound healing, and widened atrophic scars.
Marfan syndrome	*FBN1* gene/fibrillin-1 15q21.1	Unusually tall individuals; exceptionally long, thin limbs; pectus excavatum ("hollow chest"); scoliosis; ectopia lentis (dislocation of the lens); severe near-sightedness (myopia); dilatation or dissection of the aorta at the level of the sinuses of Valsalva, which may lead to cardiomyopathy or even a rupture of the aorta, dural ectasia, and mitral valve prolapse.
Neural crest cell migration		
Waardenburg syndrome	*PAX3* gene/paired box protein PAX3 2q35	Dystopia canthorum (lateral displacement of the inner canthi); growing together of eyebrows; lateral displacement of lacrimal puncta; a broad nasal root; heterochromia of the iris; congenital deafness or hearing impairment; and piebaldism, including a white forelock and a triangular area of hypopigmentation.
Congenital intestinal aganglionosis	*RET* gene/receptor tyrosine kinase 10q11.2 *GDNF* gene/glial cell-line derived neurotrophic factor 5p13.1-p14 *NRTN* gene/neurturin 19p13.3 *EDNRB* gene/endothelin B receptor 13q22 *EDN3* gene/endothelin 3 20q13.2-q13.3	Arrest of the caudal migration of neural crest cells resulting in the absence of ganglionic cells in the myenteric and submucosal plexuses; abdominal pain and distension; inability to pass meconium within the first 48 hours of life; gushing of fecal material upon a rectal digital exam; constipation; emesis; a loss of peristalsis in the colon segment distal to the normal innervated colon; and the failure of internal anal sphincter to relax following rectal distension (i.e., abnormal rectoanal reflex).
Orofacial clefting	*DLX 1-6* gene family/homeodomain gene regulatory proteins 2q32, 2cen-q33,17q21.3-q22, 17q21.33,7q22, 7q22 *SHH* gene/shh protein 7q36 *TGF-α* gene/TGF-αvariant protein 2p13 *TGF-β* gene/TGF-βprotein 14q24 *IRF-6* gene/IRF6 transcription factor 1q32.3-q41	Cleft lip (unilateral cleft lip on the left side is most commonly seen); cleft palate; a combination of both cleft lip and cleft palate; and other associated abnormalities in the central nervous system, skeletal system, cardiovascular system, and all tissue of neuroectodermal origin.
Treacher Collins syndrome	*TCOFI* gene/treacle protein 5q32-q33.1	Hypoplasia of the zygomatic bones and mandible resulting in midface hypoplasia, micrognathia, and retrognathia; external ear abnormalities including small, absent, malformed, or rotated ears; and lower eyelid abnormalities including coloboma.

X. SELECTED PHOTOGRAPHS OF DEVELOPMENTAL DISORDERS (Figure 15-2)

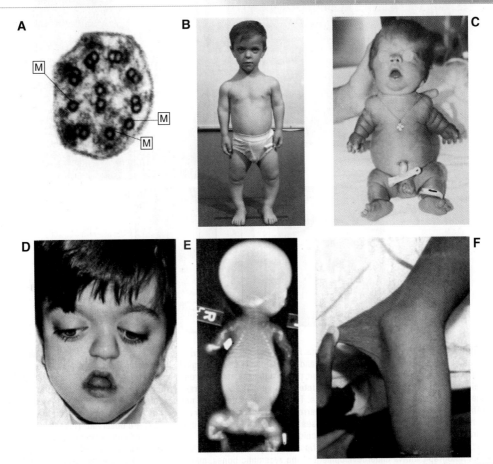

FIGURE 15-2. Birth defects associated with various developmental processes. (A) Primary ciliary dyskinesia (PCD; immotile cilia syndrome). Electron micrograph shows a cilium from an individual with PCD where the outer dynein arms are absent and with three abnormal single microtubules (M) instead of the normal 9+2 arrangement. **(B) Achondroplasia.** Photograph shows a boy with short stature, short limbs (particularly in the proximal portions), short fingers, disproportionate trunk, bowed legs, relatively large head, prominent forehead, and deep nasal ridge. **(C) Thanatophoric dysplasia.** Photographs shows a newborn infant born at 32 weeks of gestation with a depressed nasal bridge, short extremities and extra skinfold creases, small chest, and prominent abdomen. The infant died a few hours after this picture was taken. **(D)Crouzon syndrome (E) Osteogenesis imperfecta.** Radiograph shows multiple bone fractures of the upper and lower limbs resulting in an accordionlike shortening of the limbs. **(F) Ehlers-Danlos syndrome.** Photograph shows the extremely stretchable skin of the infant. *(continued)*

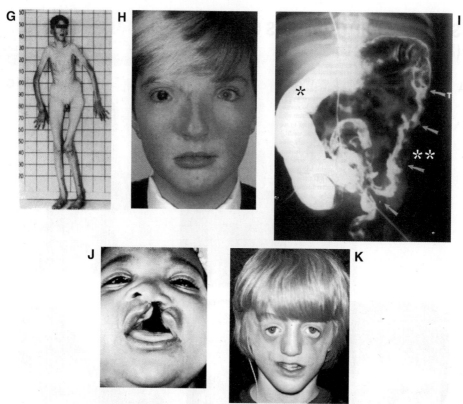

FIGURE 15-2. *(Continued)* **(G) Marfan syndrome.** Photograph shows a girl with an unusually tall stature, exceptionally long limbs, and arachnodactyly (elongated hands and feet with very slender digits). **(H) Waardenburg syndrome.** Photograph shows a young boy with a white forelock of hair, partial albinism, heterochromia of the iris, and lateral displacement of the medial canthi. **(I) Nonsyndromic congenital intestinal aganglionosis (Hirschsprung disease).** Radiograph after barium enema of a patient with Hirschsprung disease. The upper segment of the normal colon (*) is distended with fecal material. The lower segment of the colon (**) is narrow. The lower segment is the portion of the colon where the ganglionic cells in the myenteric and submucosal plexuses are absent and peculiar contractions are observed. This case shows a high transition zone (T) between the normal colon and aganglionic colon. The arrows indicate a long segment of aganglionic descending colon. **(J) Unilateral cleft lip and cleft palate. (K) Treacher Collins syndrome (mandibulofacial dysostosis).** Photograph shows underdevelopment of the zygomatic bones, mandibular hypoplasia, lower eyelid colobomas, downward-slanting palpebral fissures, and malformed external ears (note the hearing aid cord).

Review Test

1. The most common genetic cause of human birth defects is which one of the following?

(A) chromosome abnormalities
(B) multifactorial inheritance
(C) Mendelian single gene inheritance
(D) teratogen exposure

2. Which one of the following does not have a genetic component?

(A) cleft lip
(B) amniotic band
(C) spina bifida
(D) polydactyly

3. A single umbilical artery with heart defects is an example of which one of the following?

(A) a syndrome
(B) a polytopic field defect
(C) an association
(D) a sequence

4. Which one of the following establishes the anterior/posterior axis in human development?

(A) the primitive streak
(B) the notchord
(C) the future left side of the embryo
(D) the future right side of the embryo

5. The anterior/posterior location of a number of anatomical structures is determined by which of the following?

(A) the PAX gene family
(B) the HOX gene family
(C) the TGF-β gene family
(D) the fibroblast growth factor family

6. Increased paternal age is associated with a risk of having a child with which one of the following?

(A) osteogenesis imperfecta
(B) achondroplasia
(C) Ehlers Danlos syndrome
(D) Marfan syndrome

7. Which one of the following modes of inheritance is associated with orofacial clefting?

(A) autosomal recessive
(B) autosomal dominant
(C) X-linked recessive
(D) multifactorial

8. Neural crest derived cell types are regulated by which one of the following?

(A) the HOX gene family
(B) the fibroblast growth factor family
(C) the PAX gene family
(D) TGF-β family

Answers and Explanations

1. **The answer is (B).** Multifactorial inheritance accounts for approximately 25% of all human birth defects.

2. **The answer is (B).** Amniotic band is a deformation caused by mechanical forces that interfere with normal growth.

3. **The answer is (D).** A single umbilical artery is often associated with heart defects but no syndrome can be identified.

4. **The answer is (A).** The establishment of the anterior/posterior axis is associated with the appearance of the primitive streak.

5. **The answer is (B).** The HOX gene family determines the anterior/posterior orientation in the developing embryos.

6. **The answer is (B).** Many of the mutations responsible for achondroplasia are de novo and their origin is mostly paternal.

7. **The answer is (D).** Orofacial clefting is multifactorial, being determined by both genetic and environmental factors.

8. **The answer is (C).** The PAX gene family codes for DNA-binding transcription factors that regulate neural crest derived cell types.

16 Genetics of Cancer

I. THE DEVELOPMENT OF CANCER (ONCOGENESIS)

In general, cancer is caused by mutations of genes that regulate the **cell cycle, DNA repair, and/or programmed cell death (i.e., apoptosis)**. A majority of cancers (so-called **"sporadic cancers"**) are caused by mutations of these genes in somatic cells that then divide wildly and develop into a cancer. A minority of cancers (so-called **"hereditary cancers"**) are predisposed by mutations of these genes in the parental germ cells that are then passed on to their children. In addition, certain cancers are linked to **environmental factors** as prime etiological importance (e.g., bladder cancer/aniline dyes, lung cancer/smoking or asbestos, liver angiosarcoma/polyvinyl chloride, skin cancer/tar or UV irradiation). From a scientific point of view, the cause of cancer is not entirely a mystery but still remains in the theoretical arena which include the following:

A. **Standard Theory (Figure 16-1).** The standard theory suggests that cancer is the result of cumulative **mutations in proto-oncogenes and/or tumor suppressor genes**, eventually producing a cancer cell. However, if cancer is caused only by mutations in these specific cell cycle genes, it is very hard to explain the appearance of the nucleus in a cancer cell. The nucleus in a cancer cells looks as if something has detonated an explosion, resulting in an array of chromosomal aberrations (e.g., chromosome pieces, scrambled chromosomes, chromosomes fused together, wrong number of chromosomes, chromosomes with missing arms, or chromosome with extra segments; so-called **"karyotype chaos"**). The question is: "Which comes first, the mutations in cell cycle genes or the chromosomal aberrations?"

B. **Modified Standard Theory.** The modified standard theory suggests that cancer is the result of a **dramatically elevated random mutation rate** caused by environmental carcinogens or malfunction in the DNA replication machinery or DNA repair machinery. The random mutations eventually hit the proto-oncogenes and/or tumor suppressor genes, producing a cancer cell.

C. **Early Instability Theory.** The early instability theory suggests that cancer is the result of **disabling (either by mutation or epigenetically) of "master genes" that are required for cell division.** No specific master genes have been identified. Therefore, each time a cell undergoes the complex process of cell division, some daughter cells get chromosomes fused together, the wrong number of chromosomes, chromosomes with missing arms, or chromosome with extra segments, which will affect gene dosage of the proto-oncogenes and tumor-suppressor genes. The chromosomal aberrations get worse with each cell division, eventually producing a cancer cell.

D. **All-Aneuploidy Theory.** The all-aneuploidy theory suggests that cancer is the result of **aneuploidy** (i.e., abnormal number of chromosomes) that occurs during cell division. Although a

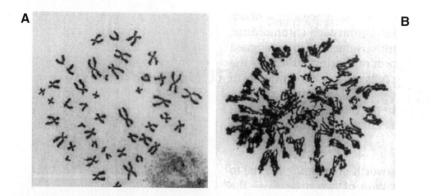

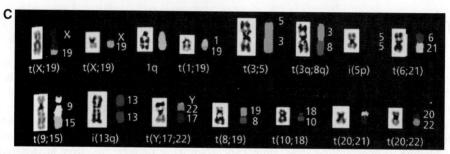

FIGURE 16-1. Karyotype chaos in a cancer cell. (A) Photograph shows a normal human karyotype. **(B)** Photograph shows an abnormal human karyotype due to a mutation involving the RAD 17 checkpoint protein, which plays a role in the cell cycle. This mutation results in a re-replication of already replicated DNA and an abnormal karyotype. **(C)** Spectral karyotyping (24-color chromosome painting) shows twelve chromosome translocations (t) and two isochromosomes in a human urinary bladder carcinoma. See Color Plate.

great majority of aneuploid cells undergo apoptosis, the few surviving cells will produce progeny that are also aneuploid. The chromosomal aberrations get worse with each cell division, eventually producing a cancer cell.

II. PHASES OF THE CELL CYCLE

The phases of the cell cycle include: G_0 **(gap) phase, G_1 (gap) phase, S (synthesis) phase, G_2 (gap) phase**, and the **M (mitosis) phase** (see Chapter 9).

III. CONTROL OF THE CELL CYCLE (Figure 16-2)

The control of the cell cycle involves three main components, which include:

A. Cdk-Cyclin Complexes. The two main protein families that control the cell cycle are **cyclins** and the **cyclin-dependent protein kinases (Cdks)**. A cyclin is a protein that regulates the activity of Cdks and is named because cyclins undergo a cycle of synthesis and degradation during the cell cycle. The cyclins and Cdks form complexes called **Cdk-cyclin complexes**. The ability of Cdks to phosphorylate target proteins is dependent on the particular cyclin that complexes with it.

1. **Cdk2-cyclin D and Cdk2-cyclin E** mediate the $G_1 \rightarrow S$ **phase** transition at the G_1 **checkpoint**.
2. **Cdk1-cyclin A and Cdk1-cyclin B** mediate the $G_2 \rightarrow M$ **phase** transition at the G_2 **checkpoint**.

B. Checkpoints. The checkpoints in the cell cycle are specialized, signaling mechanisms that regulate and coordinate the cell response to **DNA damage** and **replication fork blockage**. When the extent of DNA damage or replication fork blockage is beyond the steady-state threshold of DNA repair pathways, a checkpoint signal is produced and a checkpoint is activated. The activation of a checkpoint slows down the cell cycle so that DNA repair may occur and/or blocked replication forks can be recovered. **This prevents DNA damage from being converted into inheritable mutations producing highly transformed, metastatic cells.**

1. **Control of the G_1 checkpoint.** There are three pathways that control the G_1 checkpoint which include:

   a. Depending on the type of the DNA damage, **ATR kinase** and **ATM kinase** will activate (i.e., phosphorylate) **Chk1 kinase** or **Chk2 kinase**, respectively. The activation of Chk1 kinase or Chk2 kinase causes the inactivation of **CDC25A phosphatase**. The inactivation of CDC25A phosphatase causes the downstream stoppage at the G_1 checkpoint.

   b. Depending on the type of the DNA damage, **ATR kinase** and **ATM kinase** will activate (i.e., phosphorylate) **p53** which allows p53 to disassociate from **Mdm2**. The activation of p53 causes the transcriptional upregulation of **p21**. The binding of p21 to the Cdk2-cyclin D and Cdk2-cyclin E inhibits their action and causes downstream stoppage at the G_1 checkpoint.

   c. Depending on the type of the DNA damage, **ATR kinase** and **ATM kinase** will activate (i.e., phosphorylate) **p16** which inactivates **Cdk4/6-cyclin D** and thereby causes downstream stoppage at the G_1 checkpoint.

2. **Control of the G_2 checkpoint.** Depending on the type of the DNA damage, **ATR kinase** and **ATM kinase** will activate (i.e., phosphorylate) **Chk1 kinase** or **Chk2 kinase**, respectively. The activation of Chk1 kinase or Chk2 kinase causes the inactivation of **CDC25C phosphatase**. The inactivation of CDC25C phosphatase will cause the downstream stoppage at the G_2 checkpoint.

C. Inactivation of Cyclins. Cyclins are inactivated by **protein degradation** during **anaphase of the M phase**. The cyclin genes contain a homologous DNA sequence called a **destruction box**. A specific **recognition protein** binds to the amino acid sequence coded by the destruction box, which allows **ubiquitin** (a 76 amino acid protein) to be covalently attached to lysine residues of cyclin by the enzyme **ubiquitin ligase**. This process is called **polyubiquitination**. Polyubiquitinated cyclins are rapidly degraded by proteolytic enzyme complexes called **proteosomes**. Polyubiquitination is a widely occurring process for marking many different types of proteins (cyclins are just a specific example) for rapid degradation.

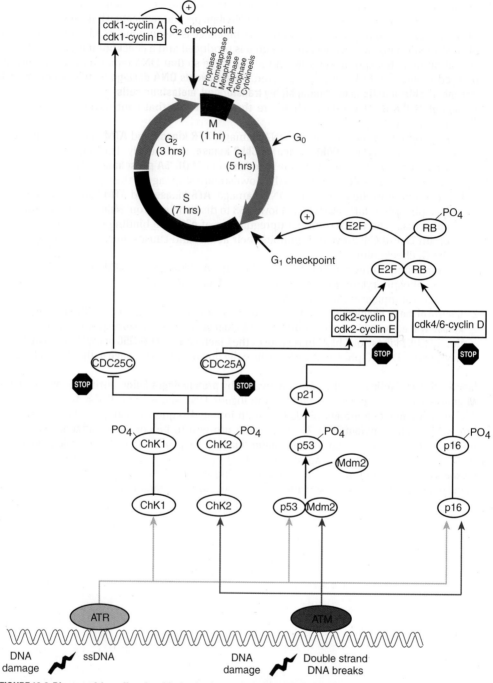

FIGURE 16-2. Diagram of the cell cycle with checkpoints and signaling mechanisms. ATR kinase responds to the sustained presence of single-stranded DNA (ssDNA) because ssDNA is generated in virtually all types of DNA damage and replication fork blockage by activation (i.e., phosphorylation) of **Chk1 kinase, p53,** and **p16**. ATM kinase responds particularly to **double-stranded DNA breaks** by activation (i.e., phosphorylation) of **Chk2 kinase, p53,** and **p16**. The downstream pathway past the STOP sign is as follows: Cdk2-cyclinD, Cdk2-cyclinE, and Cdk4/6-cyclinD phosphorylate the E2F-RB complex, which causes phosphorylated RB to disassociate from E2F. E2F is a transcription factor that causes the expression of gene products that stimulate the cell cycle. Note the location of the four stop signs. → =activation, ⊤ = inactivation.

IV. PROTO-ONCOGENES AND ONCOGENES

A. Definitions.

1. A **proto-oncogene** is a normal gene that encodes a protein involved in **stimulation of the cell cycle**. Because the cell cycle can be regulated at many different points, proto-oncogenes fall into many different classes (i.e., **growth factors, receptors, signal transducers,** and **transcription factors**).

2. An **oncogene** is a mutated proto-oncogene that encodes for an **oncoprotein** involved in the **hyperstimulation of the cell cycle,** leading to oncogenesis. This is because the mutations caused increased activity of the oncoprotein (either a hyperactive oncoprotein or increased amounts of normal protein), not a loss of activity of the oncoprotein.

B. Alteration of a Proto-Oncogene to an Oncogene.

We know now that the vast majority of human cancers are not caused by viruses. Instead, most human cancers are caused by the alteration of proto-oncogenes so that oncogenes are formed producing an oncoprotein. The mechanisms by which proto-Oncogenes are altered include:

1. **Point mutation.** A point mutation (i.e., a **gain-of-function mutation**) of a proto-oncogene leads to the formation of an oncogene. A **single mutant allele** is sufficient to change the phenotype of a cell from normal to cancerous (i.e., a **dominant mutation**). This results in a hyperactive oncoprotein that hyperstimulates the cell cycle leading to oncogenesis. Note: proto-oncogenes only require a mutation in one allele for the cell to become oncogenic, whereas tumor suppressor genes require a mutation in both alleles for the cell to become oncogenic.

2. **Translocation** (see Chapter 11). A translocation results from breakage and exchange of segments between chromosomes. This may result in the formation of an oncogene (also called a fusion gene or chimeric gene) which encodes for an oncoprotein (also called a fusion protein or chimeric protein). A good example is seen in chronic myeloid leukemia (CML). CML **t(9;22)(q34;q11)** is caused by a reciprocal translocation between chromosomes 9 and 22 with breakpoints at q34 and q11, respectively. The resulting der(22) is referred to as the **Philadelphia chromosome**. This results in a hyperactive oncoprotein that hyperstimulates the cell cycle leading to oncogenesis.

3. **Amplification.** Cancer cells may contain hundreds of extra copies of proto-oncogenes. These extra copies are found as either small paired chromatin bodies separated from the chromosomes (double minutes) or as insertions within normal chromosomes. This results in increased amounts of normal protein that hyperstimulates the cell cycle leading to oncogenesis.

4. **Translocation into a transcriptionally active region.** A translocation results from breakage and exchange of segments between chromosomes. This may result in the formation of an oncogene by placing a gene in a transcriptionally active region. A good example is seen in Burkitt Lymphoma. **Burkitt Lymphoma t(8;14)(q24;q32)** is caused by a reciprocal translocation between band q24 on chromosome 8 and band q32 on chromosome 14. This results in placing the **MYC** gene on chromosome 8q24 in close proximity to the **IGH gene** locus (i.e., an immunoglobulin gene locus) on chromosome 14q32, thereby putting the *MYC* gene in a transcriptionally active area in B lymphocytes (or antibody-producing plasma cells). This results in increased amounts of normal protein that hyperstimulates the cell cycle leading to oncogenesis.

D. Mechanism of Action of the *RAS* Gene: A Proto-Oncogene (Figure 16-3).

E. A List of Proto-Oncogenes (Table 16-1).

V. TUMOR-SUPPRESSOR GENES

A **tumor-suppressor gene** is a normal gene that encodes a protein involved in **suppression of the cell cycle**. Many human cancers are caused by **loss-of-function mutations** of tumor-suppressor genes. Note: tumor suppressor genes require a mutation in both alleles for a cell to become oncogenic, whereas proto-oncogenes only require a mutation in one allele for a cell to become oncogenic. Tumor suppressor genes can be either "gatekeepers" or "caretakers".

A. Gatekeeper Tumor Suppressor Genes. These genes encode for proteins that either regulate the transition of cells through the checkpoints ("gates") of the cell cycle or promote apoptosis. This prevents oncogenesis. Loss-of-function mutations in gatekeeper tumor-suppressor genes lead to oncogenesis.

B. Caretaker Tumor Suppressor Genes. These genes encode for proteins that either detect/repair DNA mutations or promote normal chromosomal disjunction during mitosis. This prevents oncogenesis by maintaining the integrity of the genome. Loss-of-function mutations in caretaker tumor suppressor genes lead to oncogenesis.

C. Mechanism of Action of the *RB1* Gene: A Tumor-Suppressor Gene (Retinoblastoma; Figure 16-4).

D. Mechanism of Action of the *TP53* Gene: A Tumor-Suppressor Gene ("Guardian of the Genome"; Figure 16-5).

E. A List of Tumor-Suppressor Genes (Table 16-2).

VI. HEREDITARY CANCER SYNDROMES

A. Hereditary Retinoblastoma (RB).
  1. Hereditary RB is an **autosomal dominant** genetic disorder caused by a mutation in the ***RB1* gene** on **chromosome 13q14.1-q14.2** for the ***retinoblastoma* (RB) associated protein (p110^{RB})**. >1,000 different mutations of the *RB1* gene have been identified which include missense, frameshift, and RNA splicing mutations, which result in a premature STOP codon, and a **loss-of-function mutation**.
  2. RB protein binds to **E2F** (a gene regulatory protein) such that there will be no expression of target genes whose gene products stimulate the cell cycle at the G1 checkpoint. The RB protein belongs to the family of **tumor-suppressor genes**.
  3. Hereditary RB affected individuals inherit one mutant copy of the *RB1* gene from their parents (an inherited germline mutation) followed by a somatic mutation of the second copy of the *RB1* gene later in life.
  4. **Parents of the proband.** The proband may have a RB affected parent or an unaffected parent who has an *RB1* gene mutation. If the proband mutation is identified in either parent, then the parent is at risk of transmitting that *RB1* gene mutation to other offspring. If the proband mutation is not identified in either parent, then the proband has a *de novo RB1* gene germline mutation (90% to 94% chance) or one parent is mosaic for the *RB1* gene mutation (6% to 10% chance).
  5. **Siblings of the proband.** The risk to each sibling of the proband of inheriting the *RB1* gene germline mutation is 50% if a parent has the same *RB1* gene germline mutation identified in the proband. The risk to each sibling of the proband is 3% (due to the possibility of germline mosaicism in one parent) if neither parent has the same *RB1* gene germline mutation identified in the proband.
  6. **Offspring of the proband.** The risk to the offspring of the proband is 50% if the proband has hereditary RB (bilateral tumors).

7. How can cancer due to tumor suppressor genes be autosomal dominant when both copies of the gene must be inactivated in order for tumor formation to occur? The inherited deleterious allele is in fact transmitted in an autosomal dominant manner and most heterozygotes do develop cancer. However, while the predisposition for cancer is inherited in an **autosomal dominant manner**, changes at the cellular level require the loss of both alleles, which is a **recessive mechanism**.

8. **Prevalence.** The prevalence of retinoblastoma is 1/20,000 births.

9. **Clinical features include:** a malignant tumor of the retina develops in children <5 years of age; whitish mass in the pupillary area behind the lens (leukokoria; the cat's eye; white-eye reflex); and strabismus.

B. **Classic Li-Fraumeni Syndrome (LFS).**

1. Classic LFS is an autosomal dominant genetic disorder caused by a mutation in the **TP53 gene** on **chromosome 17p13.1** for the **cellular tumor protein 53 ("the guardian of the genome")**. Mutations of the *TP53* gene have been identified which include missense (80%) and RNA splicing (20%) mutations, which result in a premature STOP codon and a **loss-of-function mutation**.

2. The activation (i.e., phosphorylation) of p53 causes the transcriptional upregulation of **p21**. The binding of p21 to the Cdk2-cyclin D and Cdk2-cyclin E inhibits their action and causes downstream stoppage at the G_1 checkpoint. p53 belongs to the family of **tumor-suppressor genes**.

3. **Parents of the proband.** Most probands have a LFS affected parent. The frequency of *de novo* mutations is not known.

4. **Siblings of the proband.** The risk to each sibling of the proband of inheriting the *TP53* gene germline mutation is 50% if a parent has the same *TP53* gene germline mutation identified in the proband. The risk to each sibling of the proband is low if neither parent has the same *TP53* gene germline mutation identified in the proband; a *de novo* mutation is assumed). No instances of germline mosaicism have been reported.

5. **Offspring of the proband.** The risk to the offspring of the proband is 50%.

6. **Prevalence.** The prevalence of LFS is 1/400 families worldwide.

7. **Clinical features include:** a highly penetrant cancer syndrome associated with soft-tissue sarcoma, breast cancer, leukemia, osteosarcoma, melanoma, and cancers of the colon, pancreas, adrenal cortex, and brain; 50% of the affected individuals develop cancer by 30 years of age and 90% by 70 years of age; an increased risk for developing multiple primary cancers; LFS is defined by: a proband with a sarcoma diagnosed <45 years of age AND a first-degree relative <45 years of age with any cancer AND a first or second-degree relative <45 years of age with any cancer.

C. **Neurofibromatosis Type 1 (NF1; von Recklinghausen disease).**

1. NF1 is a relatively common **autosomal dominant** genetic disorder caused by a mutation in the **NF1 gene** on **chromosome 17q11.2** for the **neurofibromin** protein. >500 different mutations of the *NF1* gene have been identified which include missense, nonsense, frameshift, whole gene deletions, intragenic deletions, and RNA splicing mutations; all of which result in a **loss-of-function mutation**.

2. Neurofibromin downregulates **p21 RAS oncoprotein** so that the *NF1* gene belongs to the family of **tumor-suppressor genes** and regulates cAMP levels.

3. **Parents of the proband.** ≈50% of probands have a NF1 affected parent. The other 50% of probands have an unaffected parent and develop NF1 due to a *de novo* mutation.

4. **Siblings of the proband.** The risk to each sibling of the proband of inheriting the *NF1* gene germline mutation is 50% if a parent has the same *NF1* gene germline mutation identified in the proband. The risk to each sibling of the proband is low (but > than that of the general population because of the possibility of germline mosaicism) if the parents are unaffected.

5. **Offspring of the proband.** The risk to the offspring of the proband is 50%.

6. **Prevalence.** The prevalence of NF-1 is 1/3,000 births. The *NF-1* gene has an unusually high mutation rate although the exact cause is unknown.

7. **Clinical features include:** multiple neural tumors (called **neurofibromas** that are widely dispersed over the body and reveal proliferation of all elements of a peripheral nerve including

neurites, fibroblasts, and Schwann cells of neural crest origin, numerous pigmented skin lesions (called **café au lait spots**) probably associated with melanocytes of neural crest origin, axillary and inguinal freckling, scoliosis, vertebral dysplasia, and pigmented iris hamartomas (called **Lisch nodules**).

D. Familial Adenomatous Polyposis (FAP).

1. FAP is an autosomal dominant genetic disorder caused by a mutation in the **APC gene** on **chromosome 5q21-q22** for the **a**denomatous **p**olyposis **c**oli protein. >800 different germline mutations of the *APC* gene have been identified all of which result in a **loss-of-function mutation**. The most common germline APC mutation is a **5-bp deletion** at codon 1309.

2. APC protein binds **glycogen synthase kinase 3b (GSK-3b)** which targets β-catenin. APC protein maintains normal apoptosis and inhibits cell proliferation through the **Wnt signal transduction pathway** so that *APC* gene belongs to the family of **tumor-suppressor genes**.

3. **Parents of the proband.** ≈80% of probands have a FAP affected parent. The other 20% of probands have an unaffected parent and develop FAP due to a *de novo* mutation.

4. **Siblings of the proband.** The risk to each sibling of the proband of inheriting the *APC* gene germline mutation is 50% if a parent has the same *APC* gene germline mutation identified in the proband. The risk to each sibling of the proband is low (but > than that of the general population because of the possibility of germline mosaicism) if the parents are unaffected.

5. **Offspring of the proband.** The risk to the offspring of the proband is 50%.

6. A majority of colorectal cancers develop slowly through a series of histopathological changes each of which has been associated with mutations of specific proto-oncogenes and tumor-suppressor genes as follows: normal epithelium → a small polyp involves mutation of the *APC* tumor suppressor gene; small polyp → large polyp involves mutation of *RAS* proto-oncogene; large polyp → carcinoma → metastasis involves mutation of the *DCC* tumor-suppressor gene and the *TP53* tumor-suppressor gene.

7. **Prevalence.** The prevalence of FAP is 3/100,000 individuals. FAP accounts for only 0.5% of all colorectal cancers.

8. **Clinical features include:** colorectal adenomatous polyps appear at 7 to 35 years of age, inevitably leading to colon cancer; thousands of polyps can be observed in the colon; gastric polyps may be present; and patients are often advised to undergo prophylactic colectomy early in life to avert colon cancer.

E. BRCA1 and BRCA2 Hereditary *Breast Ca*ncers.

1. BRCA1 and BRCA2 hereditary breast cancers are autosomal genetic disorders caused by a mutation in either the **BRCA1 gene** on **chromosome 17q21** for the **breast cancer type 1 susceptibility protein** or a mutation in the **BRCA2 gene** on **chromosome 13q12.3** for the **breast cancer type 2 susceptibility protein**.

2. BRCA type 1 and type 2 susceptibility proteins bind RAD51 protein, which plays a role in **double-strand DNA break repair** so that *BRCA1* and *BRCA2* genes belong to the family of **tumor-suppressor genes**.

3. >600 mutations different mutations of the *BRCA1* gene have been identified all of which result in a **loss-of-function mutation**.

4. >450 mutations different mutations of the *BRCA2* gene have been identified all of which result in a **loss-of-function mutation**.

5. **Parents of the proband.** ≈100% of individuals with a *BRCA1* or *BRCA2* gene mutation inherit the mutation from a parent. The parent may or may not have had a cancer diagnosis depending on the penetrance of the mutation, gender of the parent with the mutation, age of the parent with the mutation.

6. **Siblings of the proband.** The risk to each sibling of the proband of inheriting the *BRCA1* or *BRCA2* gene germline mutation is 50% if a parent has the same *BRCA1* or *BRCA* gene germline mutation identified in the proband. However, whether the sibling develops cancer depends on the penetrance of the mutation, gender of the sibling, and other variables.

7. **Offspring of the proband.** The risk to the offspring of the proband is 50%. However, whether the offspring develops cancer depends on the penetrance of the mutation, gender of the sibling, and other variables.

8. **Prevalence.** The prevalence of *BRCA1* gene mutations is 1/1,000 in the general population. A population study of breast cancer found a prevalence of *BRCA1* gene mutations in only 2.4% of the cases. A predisposition to breast, ovarian, and prostate cancer may be associated with mutations in the *BRCA1* gene and *BRCA2* gene although the exact percentage risk is not known and even appears to be variable within families.

9. **Clinical features include:** early onset of breast cancer, bilateral breast cancer, family history of breast or ovarian cancer consistent with autosomal dominant inheritance, and a family history of male breast cancer.

VII. LOSS OF HETEROZYGOSITY (LOH)

A. Molecular genetic analysis of RB-affected individuals revealed heterozygosity at the *RB1* gene locus in normal tissues but only one *RB1* allele in the retinoblastoma tumor tissue. That is, one *RB1* allele simply vanished in the tumor tissue and this is called LOH.

B. In other words, the tumor cells underwent LOH for a portion of chromosome 13q which included the *RB1* gene locus.

C. The remaining *RB1* allele contained the mutation and the lost *RB1* allele was normal and served as the second hit consistent with Knudson's two-hit hypothesis. In fact, LOH is the most common mechanism for the second hit in retinoblastoma. If LOH is not found, then a point mutation of the *RB1* allele is the most likely cause of the second hit.

D. The mechanisms by which LOH occurs include: deletions in chromosome 13q, mitotic nondisjunction resulting in the loss of one chromosome 13, and mitotic recombination.

E. LOH has been observed in a number of other cancers besides retinoblastoma, which include: prostate, bladder, lung, neurofibromatosis type1, familial adenomatous polyposis coli, Wilms tumor, von Hippel-Lindau, breast, and ovarian cancers.

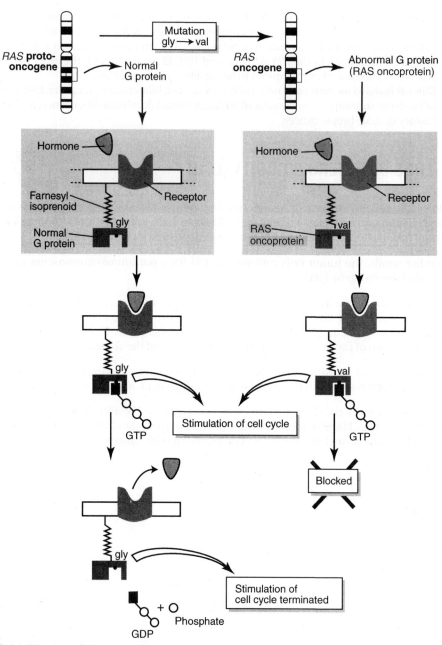

FIGURE 16-3. Diagram of *RAS* proto-oncogene and oncogene action. The *RAS* proto-oncogene encodes a normal G-protein with GTPase activity. The G protein is attached to the cytoplasmic face of the cell membrane by a lipid called farnesyl isoprenoid. When a hormone binds to its receptor, the G protein is activated. The activated G protein binds GTP, which stimulates the cell cycle. After a brief period, the activated G protein splits GTP into GDP and phosphate such that the stimulation of the cell cycle is terminated. If the *RAS* proto-oncogene undergoes a mutation, it forms the *RAS* oncogene. The RAS oncogene encodes an abnormal G protein (RAS oncoprotein) where a glycine is changed to a valine at position 12. The *RAS* oncoprotein binds GTP, which stimulates the cell cycle. However, the RAS oncoprotein **cannot** split GTP into GDP and phosphate so that the stimulation of the cell cycle is never terminated.

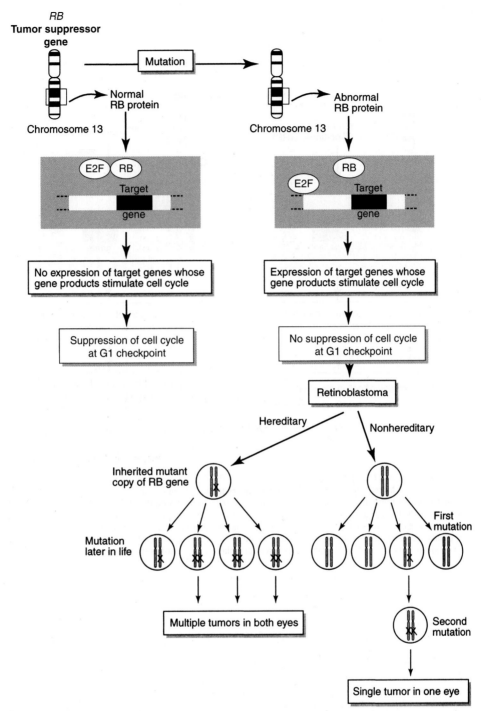

FIGURE 16-4. Diagram of *RB1* tumor-suppressor action. The *RB1* tumor-suppressor gene is located on chromosome 13q14.1 and encodes for **normal RB protein** that will bind to **E2F** (a gene regulatory protein) such that there will be no expression of target genes whose gene products stimulate the cell cycle. Therefore, there is suppression of the cell cycle at the G1 checkpoint. A mutation of the *RB1* tumor-suppressor gene will encode an **abnormal RB protein** that cannot bind E2F (a gene regulatory protein) such that there will be expression of target genes whose gene products stimulate the cell cycle. Therefore, there is no suppression of the cell cycle at the G1 checkpoint. This leads to the formation of a **retinoblastoma** tumor. There are two types of retinoblastomas. In **hereditary retinoblastoma,** the individual inherits one mutant copy of the *RB1* gene from his parents (an inherited germline mutation). A somatic mutation of the second copy of the *RB1* gene may occur later in life within many cells of the retina leading to **multiple tumors in both eyes**. In **nonhereditary retinoblastoma,** the individual does **not** inherit a mutant copy of the *RB1* gene from his parents. Instead, two subsequent somatic mutations of both copies of the *RB1* gene may occur within one cell of the retina leading to **one tumor in one eye**. This has become known as Knudson's two-hit hypothesis and serves as a model for cancers involving tumor-suppressor genes.

179

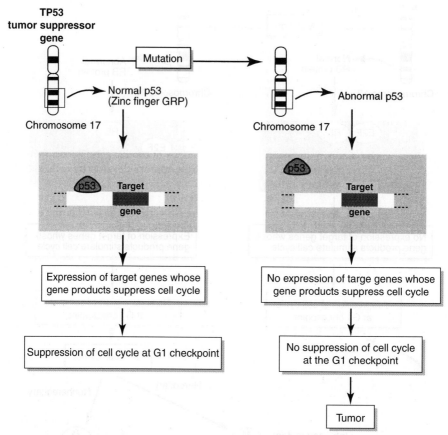

FIGURE 16-5. Diagram of *TP53* tumor-suppressor action. The *TP53* tumor-suppressor gene is located on chromosome 17p13 and encodes for **normal p53 protein (a zinc finger gene regulatory protein)** that will cause the expression of target genes whose gene products suppress the cell cycle at **G1** by inhibiting **Cdk-cyclin D** and **Cdk-cyclin E**. Therefore, there is suppression of the cell cycle at the G1 checkpoint. A mutation of *TP53* tumor-suppressor gene will encode an **abnormal p53 protein** that will cause no expression of target genes whose gene products suppress the cell cycle. Therefore, there is no suppression of the cell cycle at the G1 checkpoint. The *TP53* tumor-suppressor gene is the **most common target** for mutation in human cancers. The *TP53* tumor-suppressor gene plays a role in **Li-Fraumeni Syndrome**.

VIII. PHOTOGRAPHS OF SELECTED CANCERS (FIGURE 16-6)

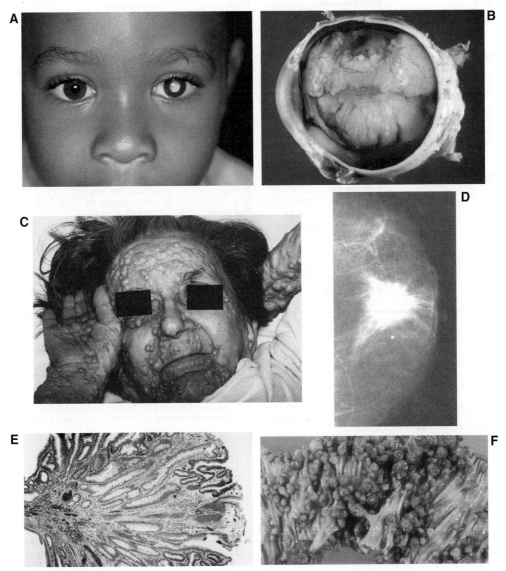

FIGURE 16-6. Cancer genetics. (A, B) Retinoblastoma (A) Photograph shows a white pupil (leukokoria; cat's eye) in the left eye. **(B)** Photograph of a surgical specimen shows the eye is almost completely filled a cream-colored intraocular retinoblastoma. **(C) Neurofibromatosis type I.** Photograph shows a woman with generalized neurofibromas on the face and arms. **(D) Breast cancer.** Mammogram shows a malignant mass that has the following characteristics: shape is irregular with many lobulations, margins are irregular or spiculated, density is medium-high, breast architecture may be distorted, and calcifications (not shows) are small, irregular, variable, and found within ducts (called ductal casts). **(E,F) Familial adenomatous polyposis. (E)** Light micrograph of an adenomatous polyp. A polyp is a tumorous mass that extends into the lumen of the colon. Note the convoluted, irregular arrangement of the intestinal glands with the basement membrane intact. **(F)** Photograph shows the colon, which contains thousands of adenomatous polyps.

table **16-1** A List of Proto-Oncogenes

Class	Protein Encoded by Proto-oncogene	Gene	Cancer Associated With Mutations of the Proto-oncogene
Growth factors	Platelet-derived growth factor (PDGF)	PDGFB	Astrocytoma, osteosarcoma
	Fibroblast growth factor	FGF4	Stomach carcinoma
Receptors	Epidermal growth factor receptor (EGFR)	EGFR	Squamous cell carcinoma of lung; breast, ovarian, and stomach cancers
	Receptor tyrosine kinase	RET	Multiple endocrine adenomatosis 2
	Receptor tyrosine kinase	MET	Hereditary papillary renal carcinoma, hepatocellular carcinoma
	Receptor tyrosine kinase	KIT	Gastrointestinal stromal tumors
	Receptor tyrosine kinase	ERBB2	Neuroblastoma, breast cancer
Signal transducers	Tyrosine kinase	ABL/BCR	CML t(9;22)(q34;q11)*
	Serine/threonine kinase	BRAF	Melanoma, colorectal cancer
	G-protein	KRAS	Lung, colon, and pancreas cancers
Transcription factors	Leucine zipper protein	FOS	Finkel-Biskes-Jinkins osteosarcoma
	Helix-loop-helix protein	N-MYC	Neuroblastoma, lung carcinoma
	Helix-loop-helix protein	MYC	Burkitt's lymphoma t(8;14)(q24;q32)
	Retinoic acid receptor (zinc finger protein)	PML/RARα	APL t(15;17)(q22;q12)
	Transcription factor	FUS/ERG	AML t(16;21)(p11;q22)
	Transcription factor	PBX/TCF3	Pre-B cell ALL t(1;19)(q21;p13.3)
	Transcription factor	FOXO4/MLL	ALL t(X;11)(q13;q23)
	Transcription factor	AFF1/MLL	ALL t(4;11)(q21;q23)
	Transcription factor	MLLT3/MLL	ALL t(9;11)(q21;q23)
	Transcription factor	MLL/MLLT1	ALL t(11;19)(q23;p13)
	Transcription factor	FLI1/EWSR1	Ewing Sarcoma t(11;22)(q24;q12)

PDGFB, platelet-derived growth factor beta gene; FGF4, fibroblast growth factor 4 gene; EGFR, epidermal growth factor receptor gene; RET, rearranged during transfection gene; MET, met proto-oncogene (hepatocyte growth factor receptor); KIT, v-kit Hardy-Zuckerman 4 feline sarcoma viral oncogene homolog; ERBB2, v-erb-b2 erythroblastic leukemia viral oncogene homolog 2; ABL/BCR, Abelson murine leukemia/Breakpoint cluster region oncogene; BRAF, v-raf murine sarcoma viral oncogene homolog B1; KRAS, Kirsten rat sarcoma 2 viral oncogene homolog; FOS, Finkel-Binkes-Jinkins osteosarcoma; N-MYC, neuroblastoma v-myc myelocytomatosis viral oncogene homolog; MYC, v-myc myelocytomatosis viral oncogene homolog; PML/RARα, promyelocytic leukemia/ retinoic acid receptor alpha; FUS/ERG, fusion (involved in t(12;16) in malignant liposarcoma)/ v-ets erythroblastosis virus E26 oncogene homolog; PBX/TCF3, pre-B-cell leukemia homeobox/ transcription factor 3 (E2A immunoglobulin enhancer binding factors E12/E47); FOXO4/MLL, forkhead box O4/myeloid/lymphoid or mixed-lineage leukemia; AFF1/MLL, AF4/FMR2 family member 1/myeloid/lymphoid or mixed-lineage leukemia, MLLT3/MLL, myeloid/lymphoid or mixed-lineage leukemia translocated to 3/myeloid/lymphoid or mixed-lineage leukemia; MLL/MLLT1, myeloid/lymphoid or mixed-lineage leukemia/myeloid/lymphoid or mixed-lineage leukemia translocated to 1; FLI1/EWSR1, Friend leukemia virus integration 1/Ewing sarcoma breakpoint region 1.
*ALL, acute lymphoblastoid leukemia; CML, chronic myeloid leukemia; APL, acute promyelocytic leukemia; AML, acute myelogenous leukemia.

table **16-2** A List of Tumor-Suppressor Genes

Class	Protein Encoded by Tumor-Suppressor Gene	Gene	Cancer Associated With Mutations of the Tumor-Suppressor Gene
Gatekeeper	Retinoblastoma associated protein p110^{RB}	RB1	Retinoblastoma, carcinomas of the breast, prostate, bladder, and lung
	Tumor protein 53	TP53	Li-Fraumeni syndrome; most human cancers
	Neurofibromin protein	NF1	Neurofibromatosis type 1, Schwannoma
	Adenomatous polyposis coli protein	APC	Familial adenomatous polyposis coli, carcinomas of the colon
	Wilms tumor protein 2	WT2	Wilms tumor (most common renal malignancy of childhood)
	Von Hippel-Lindau disease tumor suppressor protein	VHL	Von Hippel-Lindau disease, retinal and cerebellar hemangioblastomas
Caretaker	Breast cancer type 1 susceptibility protein	BRCA1	Breast and ovarian cancer
	Breast cancer type 2 susceptibility protein	BRCA2	Breast cancer
	DNA mismatch repair protein MLH1	MLH1	Hereditary nonpolyposis colon cancer*
	DNA mismatch repair protein MSH2	MSH2	Hereditary nonpolyposis colon cancer*

APC, familial adenomatous polyposis coli; VHL, von Hippel-Lindau disease; WT, Wilms tumor; NF-1, neurofibromatosis; BRCA, breast cancer; RB, retinoblastoma; TP53, tumor protein; MLH1, mut L homolog 1; MSH2, mut S homolog 2.
*See Chapter 11-II-G-5.

1. In some cases of retinoblastoma, tumors are often bilateral, their development occurs soon after birth, and other cancers, especially sarcomas, may appear later in life. In other cases, retinoblastoma displays a later age of onset and the tumors are unilateral with little increased risk of further development of cancer. What is the most likely explanation for the two different disease courses?

(A) There is a large environmental component in the bilateral cases.

(B) In the bilateral cases, the retinoblastoma proto-oncogene (Rb) is activated by a chromosome translocation.

(C) In the bilateral cases, there is already an inherited deletion of one of the Rb genes, and a second deletion or mutation occurs bilaterally.

(D) The unilateral cases are caused by the inheritance of a mutated Rb gene.

2. In familial polyposis coli, a deletion of the APC gene predisposes carriers to colon cancer. However, the cancer will not develop without the loss of the remaining APC gene. Which of the following best describes the APC gene?

(A) proto-oncogene

(B) tumor suppresser gene

(C) mitochondrial gene

(D) X-linked gene

3. In which of the following is a loss of function mutation oncogenic when it occurs in both alleles?

(A) proto-oncogene

(B) oncogene

(C) tumor suppressor gene

(D) growth factor gene

4. Which one of the following is caused by a fusion or chimeric gene created by a chromosome rearrangement?

(A) chronic myeloid leukemia

(B) retinoblastoma

(C) neurofibromatosis type I

(D) Burkitt lymphoma

5. Which of the following describes the mode of inheritance of many hereditary cancers?

(A) autosomal dominant

(B) autosomal recessive

(C) X-linked dominant

(D) X-linked recessive

6. Which of the following has been found to be the most commonly mutated gene in human cancers, most likely due to its "gatekeeper" function?

(A) the MYC proto-oncogene

(B) the IGH gene

(C) the CML fusion gene

(D) TP53 tumor suppressor gene

7. Which of the following is a likely mechanism of proto-oncogene transformation to an oncogene?

(A) loss of heterozygosity

(B) loss of function mutation

(C) chromosome translocation

(D) chromosome deletion

8. Which one of the following is hereditary cancer that is caused by a series of changes involving both tumor suppressor genes and proto-oncogenes?

(A) Li-Fraumeni syndrome

(B) von Recklinghausen disease

(C) familial adenomatous polyposis

(D) hereditary retinoblastoma

9. Which one of the following is the most likely diagnosis for a 35-year-old female patient with osteosarcoma whose mother was diagnosed with breast cancer at age 30 and who has a brother diagnosed with leukemia at age 22?

(A) Li-Fraumeni syndrome

(B) BRCA1 and BRCA2 hereditary breast cancer

(C) familial adenomatous polyposis

(D) neurofibromatosis type 1

Answers and Explanations

1. **The answer is (C).** The retinoblastoma gene is a tumor suppressor gene so both alleles of the gene must be deleted or mutated for oncogenesis to occur. Because there is already an inherited deletion or mutation present in the case of hereditary retinoblastoma, a deletion or mutation in the other allele, bilaterally, would lead to the development of bilateral tumors. Development of retinoblastoma by a series of spontaneous deletions or mutations is less likely to happen and most of the time causes only unilateral tumor development.

2. **The answer is (B).** In tumor suppressor genes, both alleles of the gene must be deleted or mutated for the tumor to occur.

3. **The answer is (C).** In tumor suppressor genes, loss-of-function mutations are oncogenic when they occur in both alleles.

4. **The answer is (A).** In chronic myeloid leukemia (CML), a balanced reciprocal translocation involving the long arms of chromosomes 9 and 22 juxtaposes the ABL and BCR proto-onco-genes, creating a fusion or chimeric gene.

5. **The answer is (A).** In most hereditary cancers, the predisposition to the disease is inherited in an autosomal dominant fashion. However, because most hereditary cancers are caused by deletions or mutations in tumor suppressor genes, both alleles must be mutated in order for a malignancy to occur. In this respect, the majority of hereditary cancers arise through a recessive mechanism.

6. **The answer is (D).** The TP53 tumor suppressor gene, often called the "guardian of the genome" has been found to be mutated in many human cancers. Most tumor suppressor genes have "gate-keeper" or "caretaker" functions.

7. **The answer is (C).** Proto-oncogenes are often activated by chromosome translocations, which juxtapose them next to other proto-oncogenes or regulatory elements as occurs with the 9:22 translocation that causes chronic myeloid leukemia (CML).

8. **The answer is (C).** The tumor suppressor genes APC, DCC, and TP53 and the proto-onco-gene RAS are all involved in the progression of colorectal cancer from normal epithelium to a carcinoma.

9. **The answer is (A).** Li-Fraumeni syndrome is suspected when there are sarcomas in a proband who is less than 45 years of age and there are any other kinds of cancers in first or second-degree relatives who are less than 45 years of age.

chapter 17 | Genetic Screening

I. PRINCIPLES OF GENETIC SCREENING

A. Genetic screening should screen for a disease that: is serious, is relatively common, and has an effective treatment.

B. Genetic screening should be relatively inexpensive, easy to perform, valid, and reliable.

C. Genetic screening should have the necessary resources for diagnosis and treatment readily available.

D. Genetic screening provides the following advantages: informed choice, improved understanding in the family affected by the genetic disorder, early treatment when available, and reduction in births of affected homozygotes.

E. A genetic screening test must have the appropriate sensitivity and specificity (Table 17-1).
1. **Sensitivity** is the ability to identify affected individuals and is measured as the **proportion of true positives.** Thus, if a test detects only 65 out of 100 affected individuals as positive (true positives; A), the test has a sensitivity of 65%. Note that 35 affected individuals are missed (false negatives; C) by the test. For example, if a patient who actually has strep throat gets a rapid strep test that comes back *positive*, then this is a true positive. However, if a patient who actually has strep throat gets a rapid strep test that comes back *negative*, then this is a false negative.
2. **Specificity** is the ability to identify unaffected individuals and is measured as the **proportion of true negatives**. Thus, if a test detects 90 out of 100 unaffected individuals as negative (true negative; D), the test has a specificity of 90%. Note that 10 unaffected individuals are falsely diagnosed as positive (false positives; B) by the test.
3. **Positive predictive value** of a screening test is the proportion of positive tests (A+B) that are true positives (A).

II. LIMITATIONS OF GENETIC SCREENING

A. Genetic screening is never 100% accurate. For example, mosaicism can confound cytogenetic results even though the accuracy of the genetic test approaches 100%. In addition, human error is always a possibility.

B. Genetic screening cannot detect the presence of disease. For example, a genetic test for hereditary breast cancer cannot predict who will get the disease.

table **17-1**	Sensitivity, Specificity, and Positive Predictive Value of a Genetic Screening Test	
	Disease Status	
Screening Test Result	**Affected Individual**	**Unaffected Individual**
POSITIVE	TRUE POSITIVE (A = 65)	FALSE POSITIVE (B = 10)
NEGATIVE	FALSE NEGATIVE (C = 35)	TRUE NEGATIVE (D = 90)

$$\text{Sensitivity} = \frac{A}{A+C} = \frac{65}{65 + 35} = 65\% = \text{Proportion of True Positives}$$

$$\text{Specificity} = \frac{D}{D+B} = \frac{90}{90 + 10} = 90\% = \text{Proportion of True Negatives}$$

$$\text{Positive Predictive Value} = \frac{A}{A+B} = 86\%$$

C. Genetic screening may not detect all the mutations causing the disease. For example, autosomal dominant breast cancer, cystic fibrosis, and Marfan syndrome all have multiple mutations, which may cause the disease. In general, it is not practical to test for all mutations.

D. Genetic screening can lead to complex ethical and social considerations, which include but are not limited to:

- Discrimination by employers of insurance companies
- Lack of effective treatment for the condition (e.g., Huntington disease, familial Alzheimer disease)
- Results may affect family members who do not wish to know about their risk for a genetic disorder
- Family members who are at risk for a genetic disorder may not wish to share this information with other family members
- "Survivor guilt," where those who do not test positive for a mutated gene may feel guilty when others in the family do test positive
- Inappropriate anxiety in carriers
- Inappropriate reassurance if test is not 100% sensitive.

III. PREIMPLANTATION GENETIC SCREENING (PGS)

The indications for PGS include:

- Couples who are at high risk of having a child with a genetic defect
- Subfertile couples
- Infertile couples

In PGS, the couple uses *in vitro* fertilization technology along with intracytoplasmic sperm injection to avoid the presence of extraneous sperm. Intracytoplasmic sperm injection is not done at every center that does PGS screening. The blastulas are biopsied at the 8-cell stage where one or two blastomeres are removed for chromosomal or DNA analysis. The 1 to 2 blastulas that are unaffected by the genetic disorder that the baby is at risk for are introduced into the uterus of the mother. Implantation must then occur for a successful pregnancy. The success rate (i.e., successful implantation/pregnancy) for this procedure is 25% per cycle of treatment.

IV. PRENATAL GENETIC SCREENING

The indications for prenatal screening include:

- Advanced maternal age
- Previous child with chromosome abnormality
- Family history of a chromosome abnormality
- Family history of a single gene disorder
- Family history of a NTD
- Family history of a congenital structural abnormality
- Consanguinity
- A poor obstetric history
- Certain maternal illnesses

Prenatal genetic screening for genetic disorders includes the following techniques:

A. **Amniocentesis.** Amniocentesis is a transabdominal sampling of **amniotic fluid** and **fetal cells.** Amniocentesis is performed at weeks 14 to 18 of gestation and is indicated in the following situations: the woman is over 35 years of age, a previous child has a chromosomal anomaly, one parent is a known carrier of a translocation or inversion, one or both parents are known carriers of an autosomal recessive or X-linked recessive trait, or there is a history of neural tube defects. The sample obtained may be used in various ways which include:

1. **α-Fetoprotein (AFP) assay** on the amniotic fluid is used to diagnose neural tube defects (NTDs); AFP is increased. Amniotic fluid levels of AFP are much more sensitive in diagnosing NTDs than maternal serum levels of AFP.
2. **Spectrophotometric assay of bilirubin** on the amniotic fluid is used to diagnose hemolytic disease of the newborn (i.e., erythroblastosis fetalis) due to Rh-incompatibility.
3. **Lecithin-sphingomyelin (L/S) ratio and phosphatidylglycerol assay** on the amniotic fluid is used to determine lung maturity of the fetus.
4. **DNA analysis on the fetal cells.** A wide variety of DNA methodologies are available to diagnose chromosomal abnormalities and single gene defects, which include: karyotype analysis, Southern blotting, or RFLP analysis (restriction fragment length polymorphism).

B. **Chorionic Villus Sampling (CVS).** CVS is a transabdominal or transcervical sampling of the chorionic villi to obtain a large amount of **fetal cells** for DNA analysis. CVS is performed at weeks 11 to 12 of gestation (i.e., much earlier than amniocentesis). The major advantage of CVS is that it offers first trimester prenatal diagnosis although it has a **1% to 2% risk of miscarriage.** The sample obtained may be used in various ways which include:

1. **Karyotype analysis**
2. **FISH analysis**
3. *M*ultiplex *l*igation-dependent *p*robe *a*mplification **(MLPA)**
4. **Biochemical assays**
5. **DNA analysis**

C. **Ultrasonography (USG).** USG is a valuable, noninvasive technique that conveys **no risk** to the mother or fetus. USG is performed at weeks 16 to 18 of gestation and is routinely offered to all pregnant women in the United States as a "dating" scan at week 12 of gestation. USG is commonly used in obstetrics to: date a pregnancy, diagnose a multiple pregnancy, assess fetal growth, determine placenta location, determine position and lie of fetus, and monitor needle or catheter insertion during amniocentesis and CVS. In addition, USG can detect certain congenital structural anomalies which include:

- **Extra digits** associated with short-limb polydactyly syndromes
- **Hypoplasia of the mandible** associated with cleft palate

- **Nuchal pad thickness** associated with Down syndrome, trisomy 13, trisomy 18, Turner syndrome, α-thalassemia, and congenital heart defects
- **Dramatic deficiency in the cranium** associated with anencephaly
- **Herniation of cerebellar tonsils** (called the **"banana sign"**) and a **distortion of the forehead** (called the **"lemon sign"**) associated with an open myelomeningocele
- **A sac in the occipital region** associated with a posterior encephalocele
- **Rocker bottom feet** associated with Trisomy 18
- **Air trapped in the stomach and duodenal bulb** associated with duodenal atresia and Down syndrome
- **Echogenic bowel** associated with meconium ileus and cystic fibrosis

D. Cordocentesis (CD). CD is the transabdominal percutaneous sampling of fetal blood from the umbilical cord vessels. The sample obtained may be used in various ways which include:
   1. Rh isoimmunization management
   2. Chromosome analysis to resolve problems associated with mosaicism found in amniocentesis or CVS.

E. Maternal Serum Screening (MSS). MSS is a sampling of the maternal blood. MSS is performed at weeks 15 to 20 of gestation. The sample obtained may be used in various ways which include:
   1. **α-Fetoprotein (AFP) assay** on the maternal serum is used to diagnose neural tube defects (NTDs); AFP is increased. However, maternal serum levels of AFP are not as sensitive in diagnosing NTDs as amniotic fluid levels of AFP. Pregnant women with maternal serum AFP level above a certain arbitrary cut-off level are offered detailed ultrasonography. Maternal AFP screening and ultrasonography have led to a striking decline in NTDs in the United States. Other causes of increased maternal serum AFP levels include: incorrect gestational age, intrauterine fetal bleed, threatened miscarriage, multiple pregnancy, congenital nephrotic syndrome, and an abdominal wall defect.
   2. **Down syndrome (Trisomy 21).** The triple test for trisomy 21 includes: **low α-fetoprotein levels (↓AFP)** in maternal serum; **low unconjugated estriol (↓estriol)** in maternal serum; and **high human chorionic gonadotropin (↑hCG)** in maternal serum. Another possible marker is **inhibin-A** which is high in maternal serum. A maternal age of >35 years of age and a positive triple test do not give an absolute diagnosis of trisomy 21, but indicate an increased probability. Pregnant women with an increased probability of carrying a trisomy 21 baby are offered amniocentesis or CVS for chromosomal analysis.
   3. **Fetal cells in the maternal circulation.** It has been demonstrated that fetal cells are present in the maternal circulation in the first trimester of pregnancy. With advances in enriching the population of fetal cells and excluding maternal cell contamination, this may provide and entirely noninvasive method for prenatal diagnosis of chromosome and DNA abnormalities in the future.

V. NEONATAL GENETIC SCREENING

The rationale for neonatal genetic screening is the prevention of subsequent morbidity in the child. In the United States, all 50 states provide neonatal genetic screening to all newborns for galactosemia, phenylketonuria, congenital hypothyroidism, and sickle cell anemia. Neonatal genetic screening is expanded to include up to 30 genetic disorders in North Carolina and Oregon.

A. Galactosemia (see Chapter 12–II-A).
   1. Neonatal genetic screening utilizes a small amount of blood obtained from a heel-prick to assay red blood cell **GALT (ga/actose-1-phosphate uridylyl/ransferase) enzyme activity** and identify GALT isoforms by **isoelectric focusing**. In addition, total red blood cell **galactose-1-phosphate** and **galactose** concentrations are measured.

2. In classic galactosemia, affected neonates have GALT enzyme activity <5% of normal values. In Duarte variant galactosemia, affected neonates have GALT enzyme activity 5% to 20% of normal values.
3. Molecular genetic testing is used clinically for: confirmatory diagnostic testing, prognostication, carrier testing, and prenatal diagnosis. Molecular genetic testing methods include:
   a. **Targeted mutation analysis.** A targeted mutation analysis panel is available for 6 common *GALT* gene mutations: Q188R, S135L, K285N, L195P, Y209C, F171S, and the Duarte variant (N314D).
   b. **Gene sequence analysis.** *GALT* gene sequence analysis is used to identify private mutations.

B. Phenylketonuria (see Chapter 12–III-A).

1. Neonatal genetic screening utilizes a small amount of blood obtained from a heel prick to assay **plasma phenylalanine** concentration using the **Guthrie card bacterial inhibition assay, fluorometric analysis,** or **tandem mass spectrometry** (this technique can be used to identify numerous other metabolic disorders on the same sample).
2. In classic phenylketonuria, affected neonates have plasma phenylalanine concentrations >16.5 mg/dL in the untreated state. In non-PKU hyperphenylalaninemia, affected neonates have plasma phenylalanine concentrations >2 mg/dL but <16.5 mg/dL when on a normal diet.
3. Molecular genetic testing is used clinically for: confirmatory diagnostic testing, prognostication, carrier testing, and prenatal diagnosis. Molecular genetic testing methods include:
   a. **Targeted mutation analysis.** A targeted mutation analysis panel is available for 15 common *PAH* gene mutations and very small deletions.
   b. **Mutation scanning.** Mutation scanning identifies virtually all point mutations in the *PAH* gene. Mutation scanning is performed by denaturing HPLC (*h*igh *p*erformance *l*iquid *c*hromatography) which is a fast and efficient method to detect locus-specific point mutations.
   c. **Gene sequence analysis.** *PAH* gene sequence analysis is used to identify private mutations.
   d. **Duplication/deletion analysis.** *M*ultiplex *l*igation-dependent *p*robe *a*mplification (MLPA) identifies large duplications and deletions when no *PAH* gene mutations have been identified by mutation scanning or gene sequence analysis.

C. Congenital Hypothyroidism.

1. Neonatal genetic screening utilizes a small amount of blood obtained from a heel prick 2 to 5 days after birth to assay **thyroxine (T$_4$)** concentration initially with a follow-up thyroid-stimulating hormone (TSH) assay if the T$_4$ value is low.
2. In congenital hypothyroidism, affected neonates have **low serum total T$_4$** and **low serum free T$_4$** concentrations along with **high serum TSH** concentrations. Neonates with TSH concentrations >60 mU/L (birth to 24 hours) or >20 mU/L (after 24 hours of birth) are recalled for evaluation.
3. **In neonates 1 to 4 days of age,** normal serum total T$_4$ concentration is 10 to 22 μg/dL and normal serum free T$_4$ is 2 to 5 ng/dL. **In neonates 1 to 4 weeks of age,** normal serum total T$_4$ concentration is 7 to 16 μg/dL and normal serum free T$_4$ is 0.8 to 2 ng/dL. Serum T$_4$ concentrations are higher in normal neonates ≈1 to 4 weeks of age versus adults due to a TSH surge soon after birth.
4. Congenital hypothyroidism is **most commonly caused (85% of cases) by thyroid dysgenesis** (e.g., agenesis, hypoplasia) during embryological development. The other 15% of cases are caused by inborn hereditary errors of thyroid hormone synthesis.
5. **Congenital hypothyroidism is the most common treatable cause of mental retardation.**
6. **Prevalence.** The prevalence of congenital hypothyroidism is ≈ 1/4,000 births.
7. **Clinical features include:** few if any clinical features are present because maternal T$_4$ crosses the placenta so that T$_4$ concentrations are ≈25% to 50% of normal; birth length and weight are normal; lethargy, slow movement, hoarse cry, feeding problems, constipation, macroglossia, umbilical hernia, large fontanels, hypotonia, dry skin, hypothermia, and jaundice may be observed.

D. **Sickle Cell Disease**.
1. Neonatal genetic screening utilizes a small amount of blood obtained from a heel prick to assay **hemoglobin S (HbS)** levels using HPLC or isoelectric focusing.
2. Neonates with hemoglobins that suggest sickle cell disease or other hemoglobinopathies are recalled for evaluation by six weeks of age.
3. Molecular genetic testing is used clinically for: confirmatory diagnostic testing, carrier testing, and prenatal diagnosis. Molecular genetic testing methods include:
    a. **Targeted mutation analysis**. PCR-based and *r*estriction *f*ragment *l*ength *p*olymorphism (RLFP) methods can be used to detect point mutations in the *HBB* gene which include: E6V mutation associated with HbS, E6K mutation associated with HbC, E121Q mutation associated with HbD, and E121K mutation associated with HbO.
    b. **Gene sequence analysis**. *HBB* gene sequence analysis is used if targeted mutation analysis is uninformative or to identify *HBB* gene mutations associated with ß-thalassemia hemoglobin variants.
4. Sickle cell disease has a very high carrier frequency in African, Mediterranean, Middle Eastern, Indian, Caribbean, and portions of Central and South American populations. In particular, sickle cell disease has a 1 in 12 carrier frequency in the African American population and a 1 in 4 carrier frequency in the west central African population. Carrier testing can be done using HPLC, isoelectric focusing, and DNA based assays.
5. Prenatal genetic testing is available by DNA analysis of fetal cells obtained by amniocentesis (at ≈15 to 18 weeks of gestation) or chorionic villus sampling (at ≈ 12 weeks of gestation). *HBB* gene mutations must be identified in both parents before prenatal testing can be done.

VI. FAMILY GENETIC SCREENING

The indications for family genetic screening are high-risk individuals or couples due to a positive family history of:

- An autosomal dominant genetic disorder with reduced penetrance or late onset (e.g., Huntington disease)
- Familial adenomatous polyposis coli
- Hereditary nonpolyposis colorectal cancer) where heterozygotes can be identified
- An autosomal recessive genetic disorder (e.g., cystic fibrosis) where heterozygotes (or carriers) can be identified
- A X-linked recessive disorder (e.g., Duchenne muscular dystrophy) where female heterozygotes (or carriers) can be identified
- Chromosomal rearrangement (e.g., translocation).

A. **Huntington Disease (HD)**.
1. Family genetic screening utilizes a commercially available kit to determine the **CAG repeat number** for each allele.
2. Asymptomatic at-risk adults seek presymptomatic family genetic screening in order to make personal decisions regarding careers, financial estates, reproductive decisions, or just the "need to know."
3. Reproductive decisions may be aided by knowledge of disease status and can provide assurance to those who do not have the mutation.
4. Genetic testing of asymptomatic at-risk adults does not accurately predict the exact age of onset, severity, type of symptoms, or the rate of progression. However, the size of the CAG trinucleotide repeat does generally correlate with age of onset.
5. Extensive pre- and posttesting counseling is required to address issues like: disability insurance coverage, employment discrimination, educational discrimination, changes family interactions, depression, and suicide ideation.
6. Genetic testing of asymptomatic at-risk individuals <18 years of age is generally not recommended.

7. Prenatal genetic testing for pregnancies at 50% risk is available by DNA analysis of fetal cells obtained by amniocentesis (at ≈15 to 18 weeks of gestation) or chorionic villus sampling (at ≈12 weeks of gestation).

B. Familial Adenomatous Polyposis Coli (FAPC).

1. Molecular genetic testing is used clinically for: confirmatory diagnostic testing, predictive testing, and prenatal diagnosis. Molecular genetic testing methods include:
   a. **Mutation scanning.** Mutation scanning identifies point mutations in the *APC* gene. Mutation scanning is performed by denaturing HPLC, which is a fast and efficient method to detect locus-specific point mutations.
   b. **Gene sequence analysis.** *APC* gene sequence analysis is used to identify mutations.
   c. **Protein truncation testing.** Protein truncation testing identifies premature truncation of the APC protein.
   d. **Duplication/deletion analysis.** MLPA identifies large duplications and deletions when no *APC* gene mutations have been identified by mutation scanning or full gene sequence analysis.
2. Genetic testing of asymptomatic at-risk adults and children can be used with certainty when a clinically diagnosed relative undergoes genetic testing and the specific mutation in the *APC* gene is identified.
3. When a clinically diagnosed relative is not available, failure to identify an *APC* gene disease-causing mutation in the at-risk adult or child does not eliminate the possibility that an *APC* gene disease-causing mutation is present in the at-risk adult or child.
4. Those with mutations in the *APC* gene can benefit from early diagnosis. With regular monitoring by colonoscopy beginning at ≈10 years of age, polyps can be removed so they cannot develop into malignancies.
5. Prenatal genetic testing for pregnancies at 50% risk is available by DNA analysis of fetal cells obtained by amniocentesis (at ≈15 to 18 weeks of gestation) or chorionic villus sampling (at ≈12 weeks of gestation). The APC disease-causing mutation in an affected family member must be identified before prenatal testing is performed.

C. Cystic Fibrosis (CF).

1. Molecular genetic testing is used clinically for: diagnosis in symptomatic individuals, carrier testing, and prenatal diagnosis. Molecular genetic testing methods include:
   a. **Targeted mutation analysis.** A targeted mutation analysis pan-ethnic panel is available for 23 common *CFTR* gene mutations.
   b. **Gene sequence analysis.** *CFTR* gene sequence analysis is used to identify mutations.
   c. **5T/TG tract analysis.** A poly T tract (a string of thymidine bases) and a TG tract (a string of TG repeats) are associated with CFTR-related disorders.
   d. **Duplication/deletion analysis.** MLPA identifies large duplications and deletions when no *CFTR* gene mutations have been identified by targeted mutation analysis or full gene sequence analysis.
2. CF has a 1 in 28 carrier frequency in Caucasian populations, 1 in 61 in the African American population, and 1 in 29 in the Ashkenazi Jewish population. Carrier testing can be done using the pan-ethnic 23-mutation panel and 5T/TG tract analysis. CF carrier testing is recommended for non-Jewish Caucasians and Ashkenazi Jews. CF carrier testing is also offered as routine prenatal care in some centers.
3. Prenatal genetic testing for pregnancies at 50% risk is available by DNA analysis of fetal cells obtained by amniocentesis (at ≈15 to 18 weeks of gestation) or chorionic villus sampling (at ≈12 weeks of gestation). *CFTR* gene mutations must be identified in both parents before prenatal testing can be done.
4. When *CFTR* gene mutations are identified in both parents, pregnancy can be achieved through assisted reproductive technology. In this case, preimplantation genetic testing is available using embryonic cells obtained from a multicell stage embryo during the *in vitro* fertilization technique.

D. Duchenne Muscular Dystrophy (DMD).

1. Molecular genetic testing is used clinically for: diagnosis in symptomatic individuals, carrier testing, and prenatal diagnosis. Molecular genetic testing methods include:

   a. **Targeted mutation analysis.** PCR-based, Southern blotting, and FISH methods can be used to detect deletions in the *DMD* gene. Southern blotting and quantitative PCR can be used to detect duplications.

   b. **Mutation scanning.** Mutation scanning identifies small deletions, small insertions, point mutations, or splicing mutations in the *DMD* gene.

   c. **Gene sequence analysis.** *DMD* gene sequence analysis also identifies small deletions, small insertions, point mutations, or splicing mutations.

   d. **Duplication/deletion analysis.** MLPA identifies large duplications and deletions in probands and carrier females.

2. DMD has a 1 in 4,000 carrier frequency in the US population although it is difficult to calculate because ≈33% of DMD cases are new mutations. When the *DMD* gene mutation of the proband is known, carrier testing can be performed by real-time PCR, FISH, or gene sequence analysis. When the *DMD* gene mutation of the proband is not known, linkage analysis can be offered to at-risk female to determine carrier status in families with more than one affected DMD male.

3. Prenatal genetic testing for pregnancies of carrier mothers is available if the *DMD* gene mutation has been identified in a family member or if linkage has been established. The usual procedure is to determine fetal sex by karyotype by DNA analysis of fetal cells obtained by amniocentesis (at ≈15 to 18 weeks of gestation) or chorionic villus sampling (at ≈12 weeks of gestation). If the karyotype is 46, XY, the DNA can be analyzed for the known *DMD* gene mutation.

VII. POPULATION GENETIC SCREENING

Population genetic screening is the systematic application of a test in a population to identify high-risk individuals who have genotypes that may lead to genetic disorders in themselves or their descendants and have not sought medical attention for the disorder. The implementation of a population genetic screening program is a very involved logistical endeavor, which requires:

- Financial, staffing, and technological resources
- A workable solution to introduce the program into the population
- A method to monitor the outcomes
- Quality control assurance

Some examples of population genetic screening are discussed below.

A. Tay-Sachs Disease.

1. Molecular genetic testing is used clinically for: confirmatory diagnosis, carrier testing, and prenatal diagnosis. Molecular genetic testing methods include:

   a. **Targeted mutation analysis.** A targeted mutation analysis panel is available for 6 common *HEXA* gene mutations: 3 null alleles (+TATC1278, +1IVC12, +1IVS9), G269S allele, and 2 pseudodeficiency alleles (R247W and R249W).

   b. **Mutation scanning.** Mutation scanning identifies *HEXA* gene mutations in individuals who are affected or have carrier-level enzyme activity but do not have a mutation identified by the above-mentioned panel.

   c. **Gene sequence analysis.** *HEXA* gene sequence analysis is used to identify *HEXA* gene mutations in individuals who are affected or have carrier-level enzyme activity but do not have a mutation identified by the above-mentioned panel.

2. Tay-Sachs disease has a 1 in 30 carrier frequency in Ashkenazi Jewish population. Carrier testing can be done by demonstrating the absence of hexosaminidase A activity (but normal

hexosaminidase B activity) using serum or white blood cells or by direct detection of *HEXA* gene mutations.

3. Options available to carrier couples are pregnancy termination and artificial insemination with noncarrier donors. In some Orthodox Jewish groups, carriers are forbidden to marry.

4. Prenatal genetic testing is available by DNA analysis or hexosaminidase A assay of fetal cells obtained by amniocentesis (at ≈15 to 18 weeks of gestation) or chorionic villus sampling (at ≈12 weeks of gestation). *HEXA* gene mutations must be identified in both parents before prenatal testing can be done.

5. Tay-Sachs disease is especially prevalent in the Ashkenazi Jewish population. In this regard, the implementation of genetic screening programs and education in the Ashkenazi Jewish population have led to a 90% decline in Tay-Sachs disease births.

B. Alpha-Thalassemia.

1. When suspicion for α-thalassemia is high, the following screening tests can be used: red blood cell indices, peripheral blood smear, red blood cell supravital staining of peripheral blood, quantitative and qualitative hemoglobin analysis.

2. Molecular genetic testing is used clinically for: diagnostic testing, carrier testing, prediction of clinical severity, and prenatal diagnosis. Molecular genetic testing methods include:

   a. **Targeted mutation analysis.** PCR-based methods can be used to detect deletions in the *HBA1* gene and *HBA2* gene. PCR primer panels that are targeted to the most common *HBA1* gene and *HBA2* gene mutations within the geographical location of the proband are generally used. Southern blotting can be used to detect less common or novel deletions.

   b. **Gene sequence analysis.** *HBA1* gene and *HBA2* gene sequence analysis is used to identify point mutations when suspicion for α-thalassemia is high and a deletion mutation is not detected.

3. α-Thalassemia has a very high carrier frequency in the African, Mediterranean, Arabic, Indian, and Southeast Asian populations. Carrier testing can be done initially by red blood cell indices, peripheral blood smear, red blood cell supravital staining of peripheral blood, quantitative and qualitative hemoglobin analysis. Suspicions can be confirmed by molecular genetic testing to detect *HBA1* gene and *HBA2* gene mutations.

4. Prenatal genetic testing is available by DNA analysis of fetal cells obtained by amniocentesis (at ≈15 to 18 weeks of gestation) or chorionic villus sampling (at ≈12 weeks of gestation). *HBA1* gene and *HBA2* gene mutations must be identified in both parents before prenatal testing can be done. If the known *HBA1* gene or *HBA2* gene mutation is present in the fetus, globin chain synthesis analysis can be performed by percutaneous umbilical blood sampling (at ≈18 to 21 weeks) especially in indeterminate-risk pregnancies. Ultrasonography demonstrating increased nuchal thickness or a large pleural effusion should lead to further prenatal genetic testing. Prenatal genetic testing is usually influenced by factors such as religion, culture, education, and number of children in the family.

5. α-Thalassemia is especially prevalent in the African, Mediterranean, Arabic, Indian, and Southeast Asian populations. In this regard, the implementation of genetic screening programs has been very effective in decreasing the incidence in α-thalassemia disease births. Screening of children, pregnant women, and individuals visiting public health facilities, premarital screening programs, and restrictions on issuance of marital certificates have also been effective.

C. Cystic Fibrosis (CF) Carrier Screening.

1. The most pressing issue in CFTR gene carrier screening by target mutation analysis is not only the >1,000 known mutations but also the differences in mutant alleles in different ethnic populations.

2. Carrier testing using the pan-ethnic panel for 23 common CFTR gene mutations identifies 80% of Caucasian carriers.

3. However, carrier testing using the pan-ethnic panel for 23 common CFTR gene mutations only identifies 70% of Hispanic carriers, 65% of African American carriers, and 50% of Asian American carriers.

4. This means that screening panels cannot identify all CFTR gene mutations even though many screening panels are designed for specific populations. Consequently, a negative screening panel for one or both of the partners does not exclude the possibility of a CF-affected offspring. This is known as the residual risk.

5. Recently developed **multiplex PCR-based testing** can identify many different mutant alleles in the CFTR gene simultaneously in a single procedure and makes it possible to establish population carrier screening for cystic fibrosis. Multiplex PCR-based testing identifies 90% of all carriers.

VI. METHODS USED FOR GENETIC TESTING

A detailed description of the methodology employed in each of the techniques listed below is beyond the scope of this book. The specific details and descriptions of each of these techniques can be easily found online using various search engines.

A. DNA Sequencing.

B. Southern Blotting.

C. Restriction Fragment Length Polymorphism (RFLP) Analysis.

D. Polyacrylamide Gel Analysis of Polymerase Chain Reaction (PCR) Products.

E. Dot Blot Hybridization Using Allele Specific Oligonucleotides.

F. Oligonucleotide Ligation Assay (OLA).

G. PCR.

H. ARMS-PCR.

I. Real Time PCR.

J. Multiplex PCR

K. Multiplex Ligation-Dependent Probe Amplification (MLPA).

L. DNA Microarrays (Chips).

M. Denaturing High Pressure Liquid Chromatography (dHPLC).

N. Conformation-Sensitive Capillary Electrophoresis (CSCE).

O. High Resolution Melt (HRM) Curve Assay.

P. Mass Spectroscopy.

Review Test

1. Which one of the following is the test with the highest sensitivity?

(A) identifies 80% of those with the disease and 95% of those without the disease
(B) identifies 90% of those with the disease and 80% of those without the disease
(C) identifies 95% of those with the disease and 70% of those without the disease
(D) identifies 99% of those with the disease and 60% of those without the disease

2. The positive predictive power of the test in question 1 is which one of the following?

(A) 65%
(B) 71%
(C) 86%
(D) 92%

3. Population genetic screening for Huntington disease does not conform to the principles of genetic screening because of which of the following?

(A) The condition cannot be effectively treated.
(B) It is an adult onset disease.
(C) Test results do not contribute to informed reproductive choices.
(D) Prenatal diagnosis is not available for this disease.

4. Which one of the following tests is offered to virtually all pregnant women in the United States?

(A) chorionic villus sampling
(B) ultrasonography
(C) amniocentesis
(D) cordocentesis

5. Which one of the following best describes neonatal genetic screening in the United States?

(A) All 50 states perform neonatal genetic screening.
(B) All 50 states perform neonatal genetic screening for the same disorders.

(C) All 50 states perform neonatal testing for cystic fibrosis.
(D) All 50 states perform neonatal genetic testing for Tay-Sachs disease.

6. An increased probability of Down syndrome is indicated by which one of the following maternal screening results?

(A) low AFP, low unconjugated estriol, and high HCG
(B) low AFP, high unconjugated estriol, and high HCG
(C) low AFP, high unconjugated estriol, and low HCG
(D) low AFP, low unconjugated estriol, low HCG

7. Genetic testing for which one of the following may result in the prevention of disease in the individual tested?

(A) Huntington disease
(B) cystic fibrosis
(C) Duchenne muscular dystrophy
(D) familial adenomatous polyposis

8. The major barrier to population cystic fibrosis carrier screening has been which one of the following?

(A) The mutations responsible for the disease have not been identified.
(B) The technology to identify mutations in the gene did not exist.
(C) Heterozygotes in various ethnic groups could not be identified.
(D) There are a large number of mutations responsible for the disease and they vary by ethnic group.

Answers and Explanations

1. **The answer is (D).** Sensitivity is the ability to identify affected individuals and is measured as the proportion of true positives. The identity of 99% of those with the disease gives this test the highest sensitivity.

2. **The answer is (B).** The sensitivity (99%) divided by the sensitivity (99%) plus the percentage of false positives (40%) is 71%.

3. **The answer is (A).** Huntington disease currently has no treatment and is invariably fatal. Test results can inform couples about their risk of having a child with Huntington disease so that informed reproductive decisions can be made. Prenatal diagnosis is available for the disease.

4. **The answer is (B).** Ultrasonography is offered to virtually all women in the United States at 12 weeks of gestation.

5. **The answer is (A).** Not all states perform the same genetic screening tests but all states perform newborn screening for galactosemia, phenylketonuria, congenital hypothyroidism, and sickle cell anemia.

6. **The answer is (A).** A low alpha-fetoprotein value, along with low unconjugated estriol and high human chorionic gonadotropin values in the maternal serum indicate an increased probability of Down syndrome in a fetus.

7. **The answer is (D).** Identification of a mutation in the APC gene would indicate that increased monitoring of the intestinal epithelium is warranted in that individual. Any polyps identified that could potentially progress to a malignancy would be removed.

8. **The answer is (D).** There are over 1,000 mutations responsible for cystic fibrosis and the ones that are responsible for most of the cases vary by ethnic group. The mutations responsible for cystic fibrosis are well characterized and new technologies may eventually allow for population screening for cystic fibrosis.

chapter 18

Consanguinity

CONSANGUINITY

Consanguinity is mating with a blood relative where there is at least one common ancestor no more remote than a great-great-grandparent. **Incest** is mating between a parent and child or between brother and sister (first-degree relatives). Consanguineous marriage is common in many parts of the world and many states within the United States allow first-cousin marriages. For example, in the Indian population, consanguineous marriage between an uncle and niece (second-degree relatives) occurs. In the Arab population, consanguineous marriage between first cousins (third-degree relatives) who are the children of two brothers occurs.

A. Consanguinity Increases the Incidence of Rare Autosomal Recessive Disorders.

B. The mating of first cousins provides the exact conditions that allow rare recessive disorders to manifest. The mating of first cousins almost doubles incidence of birth defects compared to the general population.

C. Consanguinity slightly increases the incidence of multifactorial inherited disorders.

D. Coefficient of Relationship (COR). COR is the proportion of genes in common between two related individuals. COR is described by the equation below.

$$\text{COR} = (1/2)^{n-1}$$

where n = number of
individuals in the path

E. Coefficient of Inbreeding (COI) or Homozygous by Descent. COI is the probability that an individual is homozygous at a locus as a result of consanguinity in his or her parents. COI is described by the equation below.

$$\text{COI} = (\text{COR})1/2$$

F. Summary Table of COI (Table 18-1).

G. Example of Consanguinity (Figure 18-1). Paula is a carrier of CFN (congenital Finnish nephrosis), a rare autosomal recessive kidney disease. If she mates with her first cousin Simon (a cousin through her maternal uncle), what is the chance he also carries the abnormal gene?
   1. How many individuals are in the pedigree path? Begin with Paula and ascend the pedigree path to one of the common ancestors (in this example it is Paula's grandfather). Then

t a b l e **18-1** Table of COI		

Relationship	Degree of Relationship	COR	COI
Parent-child	First degree	1/2 (50%)	1/4 (25%)
Brother-sister	First degree	1/2 (50%)	1/4 (25%)
Uncle-niece	Second degree	1/4 (25%)	1/8 (12%)
First cousins	Third degree	1/8 (12%)	1/16 (6%)
Second cousins	Fifth degree	1/32 (3%)	1/64 (1.5%)

descend the pedigree to Simon. Counting Paula and Simon, there are **5 individuals in this path** (Note the numbering on the pedigree). **Therefore n = 5.**

2. **What is the COR between Paula and Simon through the grandfather and grandmother?**
   a. Calculate the COR between Paula and Simon through the grandfather: $COR = (1/2)^{n-1} = (1/2)^{5-1} = (1/2)^4 = $ **1/16 (6%)**. Consequently, there is a 6% chance that Simon has inherited the gene if it was passed through the grandfather.
   b. Calculate the COR between Paula and Simon through the grandmother (because they also share their grandmother as a common ancestor, you must calculate the COR for her the same way): $COR = (1/2)^{n-1} = (1/2)^{5-1} = (1/2)^4 = $ **1/16 (6%)**. Consequently, there is a 6% chance that Simon has inherited the gene if it was passed through the grandmother.
   c. Calculate the final COR for first cousins: Add the common ancestors' COR's: $1/16 + 1/16 = $ **1/8 (12%)**. Therefore, the final COR for first cousins is 1/8 (12%). This can be interpreted as a 1/8 (12%) chance that Simon is a carrier of the abnormal gene.
3. **If Paula and Simon have a child that is affected with CFN, what is the chance that the child inherited both abnormal alleles from a common ancestor?** The COR for first cousins is 1/8. So, COI = (COR) 1/2 = (1/8) (1/2) = 1/16 (6%). This can be interpreted as a 1/16 (6%) chance that the CFN affected child inherited both abnormal alleles from a common ancestor.

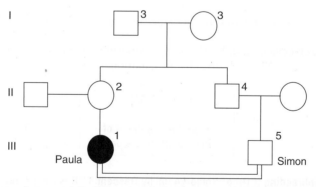

FIGURE 18-1: A pedigree of consanguinity. The numbers indicate number of individuals in the pedigree path. You can count through Paula's grandfather (3) or Paula's grandmother (3). In both cases, there are 5 individuals in the path.

Review Test

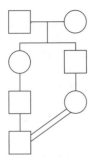

1. How is this couple related?

(A) uncle-niece
(B) first cousins
(C) first cousins once removed
(D) second cousins

2. What is the COI?

(A) 1/8
(B) 1/16
(C) 1/32
(D) 1/64

3. A marriage between first cousins would be a marriage to which one of the following?

(A) a first-degree relative
(B) a second-degree relative
(C) a third-degree relative
(D) a fourth-degree relative

4. Consanguinity increases the risk for which one of the following?

(A) autosomal recessive disorders
(B) autosomal dominant disorders
(C) mitochondrial disorders
(D) X-linked dominant disorders

5. The mating of first cousins does what to the incidence of birth defects compared to the general population?

(A) nothing
(B) doubles the risk
(C) triples the risk
(D) quadruples the risk

Answers and Explanations

1. **The answer is (C).** They are first cousins once removed because the father of the male member of the couple is the female member's first cousin.

2. **The answer is (C).** The coefficient of relationship (COR) between the couple is $1/32 + 1/32 = 1/16$ because there are 6 individuals in the path from the male member of the couple to the female member of the couple through both of the male's great grandparents. Since COR $= 1/2^{n-1}$, it is $1/2^5$ for each member of the couple or $1/32$. When the two are added, the result is $1/16$. Since the Coefficient of Inbreeding (COI) is (COR)$1/2$, the COI for this couple is $(1/16)1/2$ or $1/32$.

3. **The answer is (C).** First cousins are third-degree relatives.

4. **The answer is (A).** The risk of bringing rare recessive genes together increases when there is consanguinity because there is an increased chance that family members will share a mutated gene already present in the family.

5. **The answer is (B).** The risk for birth defects in first cousin matings has been empirically derived to be roughly double that of the general population.

Comprehensive Examination

1. Which of the following are multifactorial disorders?

(A) Marfan syndrome, neurofibromatosis
(B) Klinefelter syndrome, Turner syndrome
(C) diabetes, most cancers
(D) cystic fibrosis, sickle cell anemia

2. Which of the following is a known environmental factor in the development of a cancer?

(A) HPV
(B) BCR
(C) BRCA1
(D) Rb1

3. Certain combinations of chromosomes in Robertsonian translocations carry a risk for the carriers of these combinations to have abnormal children because the possible trisomies that can result are viable (can be born alive). Other combinations carry little or no risk for abnormal children, but can result in what is perceived as infertility because abnormal conceptions may not be recognized before they spontaneously abort. Which one of the following Robertsonian translocations in a carrier confers the greatest risk for having an abnormal child?

(A) t(14;21)
(B) t(14;15)
(C) t(15;15)
(D) t(22;22)

4. A couple has just had a third miscarriage. They have had no successful pregnancies. A review of the couple's family histories reveals that the husband's mother had two miscarriages, and the husband's sister has one child, but has had two miscarriages. The couple still wishes to have children. Which of the following is the best recommendation for how the couple should proceed?

(A) The couple should consider adoption.
(B) The couple should be tested for Fragile X.
(C) The couple should have cytogenetic studies performed.
(D) The couple should be tested for cystic fibrosis carrier status.

5. Fragile X syndrome is one of the most common causes of mental retardation in humans. It generally acts like an X-linked recessive disease, but some males do not have the disease yet they can pass it on, and some females are affected. The cause of the disease explains these observations. Fragile X syndrome is caused by which one of the following mechanisms?

(A) a deletion of the Prader-Willi/Angelman gene on the father's X chromosome
(B) a triplet repeat expansion
(C) chromosome breakage
(D) having two X chromosomes

6. Which one of the following is usually associated with remission in chronic myelogenous leukemia (CML)?

(A) acquisition of the 9;22 translocation
(B) appearance of the derivative chromosome 22 (Philadelphia chromosome)
(C) disappearance of the derivative chromosome 22 (Philadelphia chromosome)
(D) acquisition of massive hyperdiploidy

Use the following pedigree for questions 7 and 8.

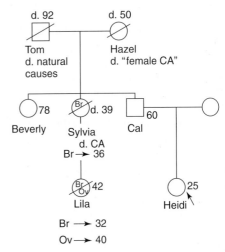

Key

Br – Breast cancer   Ov – Ovarian cancer

Heidi, age 25 and healthy, is concerned about her paternal family history of breast cancer. Her mother's side is noncontributory (no history of cancer). Heidi's paternal aunt, Sylvia, was diagnosed with breast cancer at 36 and died at 39. This aunt's daughter, Lila, was diagnosed with ovarian cancer at 40 and breast cancer at 32. She is currently in remission at 42. Heidi's paternal grandmother, Hazel, died of "female cancer" at 50. Her grandfather, Tom, died of natural causes at 92. She has one healthy aunt, Beverly, age 78. Her father, Cal, is alive and well at 65.

7. Who is the ideal person to have BRCA testing first in this family?

(A) Heidi

(B) Lila

(C) Beverly

(D) Cal

8. Regarding BRCA testing, which one of the following statements applies?

(A) Risks of BRCA testing include psychological distress, change in family dynamics, and a false sense of security if negative.

(B) If Heidi has a BRCA mutation, she has a 100% risk for breast cancer.

(C) Her only preventative option is mastectomy if she is BRCA positive.

(D) If Heidi has a BRCA mutation, each of her future children would have a 25% risk to inherit this from her.

9. Cytogenetic changes in hematological cancers, such as leukemia, generally act by which one of the following mechanisms?

(A) activation of proto-oncogenes

(B) activation of tumor suppressor genes

(C) deletion of proto-oncogenes

(D) switching off proto-oncogenes

10. Which one of the following disorders has a 100% lifetime cancer risk if untreated (100% penetrance)?

(A) FAP (familial adenomatous polyposis)

(B) HNPCC (hereditary non-polyposis colorectal cancer)

(C) hereditary breast and ovarian cancer (BRCA1 and BRCA2)

(D) hereditary melanoma

11. Which disorder conveys up to a 40% lifetime risk of ovarian cancer and a 6% risk for male breast cancer?

(A) FAP (familial adenomatous polyposis)

(B) HNPCC (hereditary non-polyposis colorectal cancer)

(C) hereditary breast and ovarian cancer (BRCA1 and BRCA2)

(D) hereditary melanoma

12. Which one of the following is an example of a good screening test?

(A) tandem mass spec testing for phenylketonuria (PKU)

(B) DNA sequence testing for BRCA1 and 2

(C) multiple mutation testing for cystic fibrosis carrier status

(D) multiple mutation testing for Marfan syndrome

13. Rami and Jillian have a son, Christian, who is newly diagnosed with autism by a developmental pediatrician. Jillian is pregnant with her second child, a male, as determined by ultrasonography. In order to determine recurrence risk, she brings the child to you, the medical geneticist. Regarding your management and counseling, which one of the following best represents your response to the couple?

(A) You counsel them that autism is always genetic and recurrence risk is 100%.

(B) You start by drawing Christian's blood for chromosome analysis and Fragile X syndrome.

(C) If chromosome analysis and Fragile X testing is normal, you tell them their child has no chance to have autism.

(D) You tell them their new child should be fine as long as he does not get any childhood vaccines or eat food with gluten, because autism is entirely due to those environmental factors.

14. The hallmarks of a hereditary cancer syndrome include which of the following?

(A) increased incidence of multiple and bilateral tumors

(B) late age of onset, i.e., postmenopausal

(C) transmission through mothers only

(D) autosomal recessive inheritance usually

15. A strong environmental component to a trait is expected when which one of the following applies?

(A) concordance is almost equal between monozygotic and dizygotic twins

(B) there is a large difference in concordance between monozygotic and dizygotic twins

(C) monozygotic concordance is close to 1.0

(D) dizygotic concordance is close to 1.0

16. Regarding type 1 and type 2 diabetes, which one of the following statements apply?

(A) Type 2 diabetes has a stronger genetic component than type 1.

(B) The recurrence risk for type 2 diabetes is minimal (<1%) for a first-degree relative; first-degree relatives would not benefit from earlier screening.

(C) Type 1 diabetes is more common than type 2 diabetes.

(D) Type 2 diabetes has a minimal environmental component.

17. Which one of the following tests provides the most benefits to a patient?

(A) testing for BRCA1 and 2 mutations

(B) Huntington disease testing for triplet repeat number

(C) testing for adenomatous polyposis coli mutations

(D) testing for cystic fibrosis carrier status

18. All 50 states have newborn screening for which of the following?

(A) PKU, galactosemia, hypothyroidism

(B) all biochemical disorders

(C) hemoglobinopathies

(D) cystic fibrosis

19. The incidence of Aarskog syndrome, an X-linked recessive disorder, is approximately 1 in 20,000 affected males. What is the frequency of carrier females?

(A) 1 in 70

(B) 1 in 140

(C) 1 in 7,000

(D) 1 in 10,000

20. When indications for Fragile X testing are present in a patient (autism, mental retardation [MR], family history of MR in males), which one of the following tests should also be done?

(A) testing for Huntington disease

(B) cytogenetic testing

(C) testing for CF carrier status

(D) testing for phenylketonuria (PKU)

21. In autosomal dominant and sex-linked diseases that are genetic lethals (individuals die before they can reproduce), the source of the disease gene in the population is which one of the following?

(A) carriers of the disease gene

(B) male carriers of the disease gene

(C) female carriers of the disease gene

(D) new mutations

22. Which one of the following diseases is autosomal recessive?

(A) Huntington disease

(B) Rett syndrome

(C) Duchenne muscular dystrophy

(D) Sickle cell disease

(E) Fragile X syndrome

23. In which disease does triplet repeat expansion occur during male gametogenesis?

(A) Huntington Disease

(B) Rett syndrome

(C) Duchenne muscular dystrophy

(D) Sickle cell disease

(E) Fragile X syndrome

24. Symptoms of this disease include a history of developmental regression, mental retardation, and characteristic hand movements in affected females.

(A) Huntington disease

(B) Rett syndrome

(C) Duchenne muscular dystrophy

(D) Sickle cell disease

(E) Fragile X syndrome

25. Symptoms in affected males with this disease include large calves, Gower maneuver, and greatly elevated levels of creatine kinase.

(A) Huntington disease

(B) Rett syndrome

(C) Duchenne muscular dystrophy

(D) Sickle cell disease

(E) Fragile X syndrome

26. Which one of the following diseases is a common cause of inherited mental retardation in males?

(A) Huntington disease

(B) Rett syndrome

(C) Duchenne muscular dystrophy

(D) Sickle cell disease

(E) Fragile X syndrome

27. Hillary comes to your preconception genetic clinic for a personal history of Tetralogy of Fallot (a conotruncal congenital heart defect). Hillary's father was born with a heart defect, has immunity problems, and schizophrenia. Hillary's brother has cleft palate and a heart defect as well. Hillary has tested positive for the microdeletion of 22q11.2 by FISH (DiGeorge syndrome). What is the best estimate of the recurrence risk for a future pregnancy?

(A) 2–3% (multifactorial risk of recurrence)
(B) 25%
(C) 50%
(D) 100%

The following pedigree applies to next question.

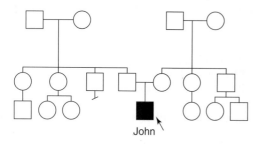

John

28. Baby John has been diagnosed with multiple congenital anomalies shortly after birth and there is no prior family history of abnormalities. Of the following, which one is the most likely explanation for the clinical findings?

(A) autosomal dominant disease
(B) X-linked dominant disease
(C) chromosome abnormality
(D) mitochondrial disease

29. Which one of the following is NOT a syndrome for which fluorescent in situ hybridization (FISH) is the primary test?

(A) Williams syndrome
(B) Prader-Willi syndrome
(C) Di George syndrome
(D) Cri-du-chat syndrome

30. Regarding the classic autosomal recessive conditions, which of the following statements applies?

(A) Pregnant women with PKU should not stay on their low phenylalanine diet during the pregnancy.
(B) People with sickle cell trait are sickly and need frequent blood transfusions.

(C) A negative cystic fibrosis carrier screen lowers a person's carrier risk to zero.
(D) Tay-Sachs disease causes progressive neurologic and physical deterioration until death around 5 years of age.

The following table applies to the next question.

The incidence of isolated sacral appendages, a multifactorial disorder, varies worldwide and is more common in North Carolina. This table gives approximate recurrence risks (%) in relation to population incidence.

Population Incidence

	France	United States	North Carolina
Individual Affected	1/1,000	1/500	1/200
One sibling	2%	3%	5%
Two siblings	8%	10%	12%
One second-degree relative	1%	1%	2%
One third-degree relative	0.5%	0.5%–1%	1%

31. In your North Carolina genetics clinic, a couple comes in for preconception counseling. Their first child had a sacral appendage and no other birth defects. What is the recurrence risk for their next child to have this birth defect?

(A) 5%
(B) 3%
(C) 2%
(D) 1%

32. Iris and Toby are considering pregnancy for the fifth time. They already have four girls, all healthy and normal appearing. Toby and his mother have Crouzon syndrome, an autosomal dominant form of craniosynostosis (premature fusion of the skull bones) with a penetrance of 100%. What is the risk for the fifth child to be affected?

(A) 0%
(B) 25%
(C) 50%
(D) 100%

The following pedigree applies to questions 33–35.

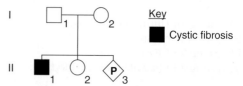

Key

■ Cystic fibrosis

33. What is I-1's risk to be a carrier?

(A) 1 (100%)
(B) 2/3 (66%)
(C) 1/2 (50%)
(D) 1/4 (25%)
(E) 0

34. What is II-2's risk to be a carrier?

(A) 1 (100%)
(B) 2/3 (66%)
(C) 1/2 (50%)
(D) 1/4 (25%)
(E) 0

35. What is the chance for their next child (II-3) to be affected?

(A) 1 (100%)
(B) 2/3 (66%)
(C) 1/2 (50%)
(D) 1/4 (25%)
(E) 0

36. Which of the following statements applies to the use of ultrasonography in pregnancy?

(A) Level II ultrasound is optimally done between 10–12 weeks of pregnancy.
(B) All babies with Down syndrome will show the "banana sign" and the "lemon sign."
(C) Ultrasonography detects 95% of major structural abnormalities but can miss small defects.
(D) Ultrasonography has no risk to mother or baby.

37. Which one of the following statements regarding prenatal diagnosis is accurate?

(A) Diagnostic testing like CVS or amniocentesis is the definitive way to diagnose a chromosome problem prior to birth.
(B) One of the benefits of CVS or amniocentesis is that it can be used as a screening test to determine the predisposition for multifactorial diseases of adulthood.
(C) One of the major advantages of amniocentesis over CVS is that it can be performed in the first trimester, although there is a higher risk of miscarriage with amniocentesis.
(D) CVS screens for open neural tube defects but amniocentesis does not.

The following is a pedigree for a family with the darkened individuals being affected with hemochromatosis:

38. What is the likelihood that the mother is a carrier of a hemochromatosis mutation?

(A) 0%
(B) 25%
(C) 50%
(D) 100%

39. Which if the following would be an appropriate indication for offering prenatal diagnosis?

(A) A woman who will be 40 years old at delivery.
(B) A fetus determined to be male by ultrasonography.
(C) A woman with a fourth-degree relative with cleft lip and palate.
(D) A woman who is pregnant by her fifth cousin.

The following pedigree applies to questions 40–42.

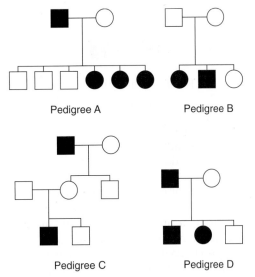

Pedigree A   Pedigree B

Pedigree C   Pedigree D

(A) pedigree A
(B) pedigree B
(C) pedigree C
(D) pedigree D

40. Which pedigree best represents the mode of inheritance for an X-linked dominant condition?

41. Which pedigree best represents the mode of inheritance for an X-linked recessive condition?

42. Which pedigree best represents the mode of inheritance for an autosomal recessive condition?

The following pedigree applies to the next question.

43. What is the most likely mode of inheritance?

(A) autosomal dominant
(B) autosomal recessive
(C) X-linked recessive
(D) X-linked dominant

The following pedigree applies to the next question.

44. What is the least likely mode of inheritance?

(A) autosomal dominant
(B) X-linked recessive
(C) mitochondrial
(D) X-linked dominant

45. Which one of the following is an accurate statement?

(A) Germline mosaicism can explain the rare occurrence of multiple affected siblings born to unaffected parents where one parent carries the disease-causing mutation only in their gametes.
(B) The same disease phenotype can be caused by mutations in different genes at different places in the genome, which is called allelic heterogeneity.

(C) A person with a genetic disease showing more severe symptoms than their affected sibling is an example of penetrance.
(D) Different mutations in the cystic fibrosis gene causing varying severity of the disease would be an example of locus heterogeneity.

For questions 46–48, match the diagnostic procedure with the appropriate description.

(A) done at 10–12 weeks, 1%–2% risk of complication
(B) done at 15–18 weeks, 0.5% risk of complication
(C) Rh isoimmunization management
(D) done at 5–6 weeks, 15% risk of complication

46. Amniocentesis

47. Cordocentesis

48. CVS

49. Which of the following statements about prenatal diagnosis is accurate?

(A) CVS should be offered if there is an elevated maternal serum AFP.
(B) Screening tests are useful to screen at risk populations and eliminate their risk if the screen is negative.
(C) Amniocentesis is recommended if there is a history of neural tube defects in a family.
(D) Cordocentesis is useful for assessing hematologic status or infection in the fetus, but the cells cannot be used for fetal karyotyping.

50. Thus far, two genes have been found that can cause autosomal dominant breast cancer (one on chromosome 13 and one on chromosome 17). This is best described as an example of which of the following?

(A) linkage
(B) allelic heterogeneity
(C) synteny
(D) linkage disequilibrium
(E) locus heterogeneity

51. An infant who appears normal until 6 months of age but then develops symptoms of progressive weakness and loss of motor skills with decreased attentiveness and an increased startled response most likely has which one of the following?

(A) metabolic disorder involving an amino acid pathway

(B) metabolic disorder involving a degradation pathway

(C) metabolic disorder involving a carbohydrate pathway

(D) metabolic disorder involving glycogen storage

52. Which one of the following clinical signs is not indicative of an inborn error of metabolism?

(A) metabolic acidosis

(B) unusual odor in urine/sweat

(C) vomiting and lethargy in an infant

(D) accelerated growth in a child

(E) mental retardation

53. A 2-week-old baby is admitted to the emergency room with a history of vomiting, no interest in feeding, and lethargy. She had been an 8-pound term baby with no neonatal problems. Which one of the following laboratory studies is not likely to help determine the diagnosis?

(A) blood glucose

(B) blood gas

(C) chromosome karyotype

(D) plasma ammonium

(E) liver function studies (bilirubin, AST, ALT)

54. Besides microdeletions, which one of the following mechanisms is known to cause Prader-Willi syndrome?

(A) chromosome duplication

(B) translocation

(C) uniparental disomy

(D) autosomal trisomy

(E) autosomal monosomy

55. It has been suggested that schizophrenia is more severe and has an earlier age of onset in the more recent generations of schizophrenia families. If so, this would be an example of which one of the following?

(A) the two-hit model

(B) imprinting

(C) delayed age of onset

(D) anticipation

(E) increased numbers of mutagens in more recent generations

56. Of the following, which statement is most applicable to the group of diseases described as "cancers"?

(A) All cancers are inherited.

(B) Most cancers are caused by environmental effects.

(C) Most cancers are caused by chromosome translocations.

(D) All cancers have a genetic component.

57. The mode of action of inherited cancers such as retinoblastoma is best described as which one of the following?

(A) autosomal recessive

(B) autosomal dominant

(C) X-linked recessive

(D) X-linked dominant

58. You examine a child in the newborn nursery whom you suspect has trisomy 21. Which one of the following tests would you order to confirm the diagnosis?

(A) maternal serum AFP

(B) cytogenetic analysis

(C) DNA sequencing

(D) neonatal genetic screening

59. Which one of the following statements is most applicable to the human genome?

(A) Humans have more chromosomes than any other organism examined to date.

(B) The human genome is composed of 100,000 genes and that is the highest number of genes seen in any organism examined to date.

(C) The human genome is composed of the greatest amount of non-coding DNA of any organism examined to date.

(D) The human genome has approximately the same number of genes as the roundworm, *Caenorhabditis elegans*.

60. The town of Lake Maracaibo, Venezuela was settled by a few families. There are a larger proportion of people affected by Huntington disease in the town than in the general population. One possible explanation for this is which of the following?

(A) selection against heterozygotes

(B) the founder effect, or genetic drift

(C) something in the water

(D) selection against homozygotes

61. The incidence of hemophilia B, an X-linked recessive disorder, is approximately 1 in 20,000 males. What is the frequency of the disease gene?

(A) 0.01

(B) 0.002

(C) 0.0004

(D) 0.00005

62. A couple comes to see you because the wife has been found to be a carrier of a balanced reciprocal translocation between the long arms of chromosomes 4 and 10. They have no children and have been trying for 15 years. They want to know what this cytogenetic diagnosis means for their hope of having children. Which one of the following best summarizes what you will tell them?

(A) They can never have children.

(B) They are at risk for infertility and that is most likely the cause of their lack of success in conceiving.

(C) Any children they have will be abnormal.

(D) The translocation is clinically irrelevant.

63. A discrete group of birth defects that are the result of a single cause is known as which one of the following?

(A) an association

(B) a deformation

(C) a syndrome

(D) a disruption

64. The causation of three forms of dwarfism, a craniofacial syndrome, and isolated craniosynostosis by mutations in the FGFR3 gene is known as which one of the following?

(A) allelic heterogeneity

(B) a compatible pathogenic mechanism of teratogenesis

(C) a defect in hypertrophy of existing cells and tissues

(D) an extrinsic developmental defect

65. You request a cytogenetic study on a patient who you suspect has Prader-Willi syndrome. The lab report comes back stating that FISH with DNA probes specific to the Prader-Willi/Angelman locus have revealed that the region is deleted in your patient. You know that the deletion is thus on which one of the following?

(A) the maternal chromosome 15

(B) the paternal chromosome 15

(C) the maternal chromosome 22

(D) the paternal chromosome 22

66. A pregnant female comes to your clinic because of a family history of Duchenne muscular dystrophy (DMD) in her maternal uncle. He was the only person affected in her family. Your patient is an only child and this is her first pregnancy. What is her chance to have a fetus that is affected with DMD?

(A) 1/8

(B) 1/12

(C) 1/16

(D) 1/24

67. Which of the following statements is accurate in regards to prenatal diagnosis?

(A) Low AFP, high HCG, and low unconjugated estriol (uE3) on a second trimester triple screen indicates an increased risk for fetal Down syndrome.

(B) Chorionic villus sampling is less risky and more accurate for fetal chromosome abnormalities than amniocentesis.

(C) First trimester maternal serum AFP screening is more accurate for detecting fetal trisomy 21 than second trimester amniocentesis with fetal karyotyping.

(D) It is standard of care to offer pregnant women of any age CVS.

68. Although conventional cytogenetic studies should always be performed, fluorescent in situ hybridization (FISH) with the appropriate DNA probe is the definitive test for which of the following?

(A) Klinefelter and Turner syndromes

(B) Williams and DiGeorge syndromes

(C) myotonic dystrophy and Fragile X syndromes

(D) neurofibromatosis and spina bifida

69. Which one of the following statements regarding neural tube defects is most applicable?

(A) Anencephaly is the most severe form of neural tube defect.

(B) A low maternal serum AFP can indicate an increased risk for a fetal open neural tube defect.

(C) Isolated neural tube defects usually follow an X-linked recessive pattern of inheritance.

(D) The prognosis for a neural tube defect depends on the age of the mother at conception.

70. The main risk for children born to inversion carriers is the chance of which of the following?

(A) Down syndrome

(B) duplications or deletions

(C) chronic myelogenous leukemia

(D) Robertsonian translocations

71. Which one of the following statements is applicable to a 45,X karyotype?

(A) Patients with a 45,X karyotype have a male phenotype.
(B) Patients with a 45,X karyotype are severely mentally retarded.
(C) Most conceptions with a 45,X karyotype spontaneously abort.
(D) A 45,X karyotype is associated with Down syndrome.

72. Mutations in different autosomal recessively inherited genes may result in leukemia in individuals with Fanconi anemia. This is an example of which of the following?

(A) locus heterogeneity
(B) allelic heterogeneity
(C) genotype-phenotype correlation
(D) de novo mutations
(E) variable expressivity

73. The balanced reciprocal translocation between chromosomes 9 and 22 with breakpoints at q34 and q11.2, respectively, is the leukemogenic event in the chronic myelogenous leukemia (CML). Specifically, this translocation causes leukemia by which of the following mechanisms?

(A) fusing proto-oncogenes and thus activating them
(B) disrupting a tumor suppressor gene in the breakpoint cluster region (BCR)
(C) deleting the proto-oncogenes ABL and BCR
(D) deleting tumor suppressor genes

74. Which of the following is a second-degree relative?

(A) great-grandchild
(B) brother
(C) maternal cousin
(D) paternal second cousin
(E) half-brother

75. Mrs. X comes to you for genetic counseling. She is pregnant with her 2nd child, a cytogenetically normal male by amniocentesis. Her mother had 2 brothers with hemophilia A and one normal brother. The patient has no brothers. What is the risk for this male fetus to have hemophilia A?

(A) 1/4
(B) 1/8
(C) 1/16
(D) 1/24

76. A woman comes to genetic clinic because her MSAFP triple screen is positive for Down syndrome. What diagnostic test would be the most appropriate to offer her?

(A) chorionic villus sampling (CVS)
(B) amniocentesis
(C) level II ultrasound
(D) preimplantation genetic diagnosis (PGD)

77. Regarding autosomal dominant susceptibility syndromes: if a family member with a known mutation carrier has a 12.5% chance of inheriting the mutation, then this relative is which one of the following?

(A) first-degree relative
(B) second-degree relative
(C) third-degree relative
(D) fourth-degree relative
(E) fifth-degree relative

78. What is the chance that the fetus in this family is affected with this X-linked recessive disease? The mother is a known carrier for hemophilia A.

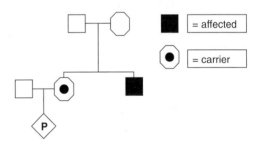

(A) 1
(B) 1/2
(C) 1/4
(D) 2/3

79. TPMT is an enzyme involved in the metabolism of thiopurine drugs. 1 in 300 (0.003) individuals are homozygous for a null mutation which causes them to have severe adverse reactions when they are given this class of drugs. If you are offered a test which has a sensitivity of 95% and a specificity of 99% to detect the individuals with a null mutation, in a population of 20,000 people, how many people would be really affected and how many people would the test call positive?

(A) 60 really affected and 257 reported positive
(B) 60 really affected and 116 reported positive
(C) 60 really affected and 57 reported positive
(D) 60 really affected and 60 reported positive

80. Maternal serum alpha-fetoprotein screening is routinely offered to pregnant women in the second trimester. Which of the following options is an appropriate action for follow-up?

(A) Offering chorionic villus sampling to a woman whose maternal serum AFP (MSAFP) shows an increased risk for Down syndrome.

(B) Recalculating the screening because the ultrasound altered the dating by 7 days.

(C) Offering amniocentesis with ultrasonography for a woman whose MSAFP indicates an increased risk for open neural tube defects.

(D) Deferring amniocentesis for a woman who will be 37 at delivery because her MSAFP was negative, which eliminates her risk.

81. There are several clinical features used to identify kindreds with familial cancer syndromes. Which of the following statements is most predictive of familial risk?

(A) environmental exposures, such as smoking, in family members with cancer

(B) your patient's current age

(C) paternal history of cancer

(D) presence of other genetic disorders in the same family

(E) proband with multiple cancers

82. The frequency of an allele for an autosomal dominant disorder in a population is 0.0001. What is the frequency of the disorder in the population?

(A) 0.01
(B) 0.02
(C) 0.0001
(D) 0.0002

83. One of your patients, a 25-year-old woman with no health problems, has just learned that her sister is the carrier of a 14;14 Robertsonian translocation. The patient desires testing and wishes to know what it will mean to her future plans of having children if she is also found to be a carrier. You tell her:

(A) She has a 15% risk of having a child with trisomy 14.

(B) She has a 5% risk of having a child with monosomy 14.

(C) She is at risk for infertility.

(D) The translocation will have no effect on her ability to have children.

84. The risk that a father carrying a mitochondrial gene mutation will pass the mutation to his son is:

(A) 100%
(B) 50%
(C) 25%
(D) 0%

85. In a family where there is bilateral cleft lip and palate, the recurrence risk would be (fill in the word or phrase from choices below) in a family where there is cleft palate only.

(A) less than
(B) greater than
(C) the same as
(D) 60% greater than

86. Sally is a carrier of the most common mutation found in the autosomal recessive disease, cystic fibrosis (CF), which is the delta F508 mutation on the long arm of chromosome 7. Sally and Ted's first child is found to have CF shortly after birth. Ted is tested and found to also be a carrier of this mutation. What is the risk of having another child with CF?

(A) 10%
(B) 25%
(C) 50%
(D) 100%

87. In a chromosome inversion, what can be said about the genes on the inverted piece of the chromosome?

(A) they are all autosomal dominant genes

(B) the genes are in reverse order on the chromosome

(C) the genes have been translocated to another chromosome

(D) they form a ring chromosome

88. Which one of the following is the karyotype of a patient with a normal phenotype?

(A) 45,X

(B) 46,XY, with a Robertsonian translocation between chromosomes 13 and 21

(C) 47,XXY

(D) 45,XX, with a Robertsonian translocation between chromosomes 13 and 21

89. Which of the following is the best estimate of the chance that a child produced by the union of a female carrier of a 21;21 Robertsonian translocation carrier and a karyotypically normal male will have Down syndrome?

(A) 0%
(B) 5%
(C) 15%
(D) 100%

90. Which of the following are examples of multifactorial disorders?

(A) Robertsonian translocations
(B) cystic fibrosis and sickle cell anemia
(C) heart disease and diabetes
(D) Duchenne muscular dystrophy and hemophilia

91. Assuming that one X chromosome in each of the following karyotypes has a mutation that causes Duchenne muscular dystrophy, which one of the karyotypes is that of an individual afflicted with this X-linked recessive disease?

(A) 47,XXY
(B) 47,XXX
(C) 46,XX
(D) 47,XYY

92. Which one of the following suggests that a disease under consideration is probably multifactorial?

(A) The disease occurs more frequently in females than males.
(B) The risk of the disease is the same for both sexes.
(C) The risk of the disease is 50% for children of an affected parent.
(D) The disease occurs in all children of an affected female.

93. Marfan syndrome is an autosomal dominant disorder. Jim has no signs of Marfan syndrome, but his mother and his son are both affected. This is an example of which of the following?

(A) allelic heterogeneity
(B) variable expressivity
(C) non-penetrance
(D) locus heterogeneity

94. Transcription factors that control spatial patterning in early human development belong to what family of genes?

(A) the abl proto-oncogene family
(B) the bcr proto-gene family.
(C) the delta F508 gene family
(D) the homeotic (HOX) gene family

95. Which one of the following should be offered amniocentesis for chromosome analysis?

(A) a woman over 35
(B) a 25-year-old woman who has had one miscarriage that was 46,XY
(C) a 25-year-old woman whose husband is 35
(D) a woman who is a carrier of the deltaF508 cystic fibrosis deletion

96. The central dogma of genetics is that a DNA code is translated into a functional protein. The correct order of events by which this is accomplished is which of the following?

(A) translation, translocation, synthesis
(B) gene, transcription, translation, protein
(C) gene, protein, transcription, translation
(D) gene, protein, translation, transcription

97. Female carriers of X-linked recessive diseases sometimes exhibit some symptoms of the disease. This is most likely due to which one of the following?

(A) variable expressivity
(B) mitochondrial inheritance
(C) Lyonization (inactivation of one of the X chromosomes in females)
(D) non-penetrance

98. In achondroplastic dwarfism, an autosomal dominant disease, the gene is lethal in the homozygous state. What is the percent of conceptions from a couple, both of whom have achondroplastic dwarfism, that will be homozygous for the gene?

(A) 0%
(B) 25%
(C) 50%
(D) 100%

99. In a pericentric inversion, which one of the following changes its position on the chromosome?

(A) the centromere
(B) the mitochondria
(C) the long arm only
(D) the short arm only

100. Sickle cell heterozygotes have some degree of protection against malaria. Because of this, the frequency of the "S" allele (the mutated allele) in sickle cell would be expected to do which one of the following?

(A) decrease in the population
(B) increase in the population

(C) not change

(D) fall to "0"

101. Because of the risk of strokes, a stat cytogenetics study is required in suspected cases of which one of the following?

(A) chronic myeloid leukemia (CML)

(B) acute promyelocytic leukemia (PML)

(C) acute myeloid leukemia (AML) with MLL translocations

(D) AML with inversion 16 and eosinophilia

102. Regarding meiosis, which of the following statements applies?

(A) In females, meiosis is suspended in prophase I until ovulation.

(B) In females, two mature ova are produced in every meiosis.

(C) At the end of meiosis, there are four mature ova.

(D) "Crossing over" does not occur during oogenesis.

103. A non-disjunction of chromosome 21 occurs during meiosis I in a woman. The non-disjoined chromosome 21 homologues go to the polar body. Which one of the following is the most likely outcome?

(A) If the ovum is fertilized by a normal sperm, the conceptus will be trisomic for chromosome 21.

(B) If the ovum is fertilized by a normal sperm, the conceptus will be monosomic for chromosome 21.

(C) If the ovum is fertilized by a normal sperm, there will be no copies of chromosome 21.

(D) If the ovum is fertilized by a normal sperm, the conceptus will have a normal karyotype.

Questions 104 and 105 are based on the following family history.

Mr. and Mrs. C have one normal female child. They have a second male child who has multiple congenital anomalies. Cytogenetic studies of the child at birth revealed a derivative chromosome 11 formed from a translocation between one chromosome 11 and one chromosome 22. The parents refused cytogenetic studies. When Mrs. C became pregnant, she refused prenatal diagnosis. The full-term male infant was born with birth defects similar to his brother's and was found by cytoge-

netic studies to have the same derivative chromosome 11. There is no family history of birth defects on either side.

104. Which one of the following is the most likely explanation for this family history?

(A) One of the parents is a carrier of a balanced reciprocal translocation between chromosomes 11 and 22.

(B) Mrs. C carries an X-linked recessive gene.

(C) Mr. C is the carrier of a Robertsonian translocation.

(D) The translocations are a coincidence.

105. Which one of the following statements about the above family history is true?

(A) The derivative chromosome in the brothers came from adjacent segregation of the balanced reciprocal translocation at meiosis in the carrier parent.

(B) The derivative chromosomes in the brothers are the result of two unrelated events.

(C) The sister probably has the same karyotype as her brothers.

(D) The birth defects in the brothers are most likely the result of X-linked recessive inheritance.

106. The genotype of a person for any specific gene is determined by which one of the following?

(A) affinity of the gene for transcription factor binding motifs

(B) the number of introns the gene contains

(C) the presence or absence of short interspersed elements within the gene

(D) the alleles present at the gene locus

107. The functional portion of the gene that codes for proteins is which one of the following?

(A) introns

(B) exons

(C) promoter

(D) enhancer

108. A pregnant patient with incontinentia pigmenti (X-linked dominant condition) comes to you for her obstetrical care and she gives you the following history: She has had 2 male stillbirths, one early miscarriage (sex unknown), a daughter who is healthy with no physical or mental abnormalities, and a daughter who has seizures, slow learning, abnormal skin pigmentation, and cone-shaped teeth. The patient has amniocentesis, which demonstrates the fetus is male. When you counsel the patient, which of the following will you discuss with her?

(A) The risk for the male fetus to be affected is 100%.
(B) The risk for her healthy daughter to have an affected son is 100%.
(C) The risk for the male fetus to be affected is 50%.
(D) X-linked dominant conditions are usually lethal in females.

109. Rudy is affected with Gaucher disease, an autosomal recessive condition. His wife is not affected but is a carrier. What is the risk for their daughter to be affected with the condition?

(A) 0%
(B) 25%
(C) 50%
(D) 75%

Questions 110–112 are based on the following family history.

Mrs. H and all three of her children, two sons and a daughter, have been diagnosed with the same genetic disease. Her daughter's son and daughter are both affected. One of her sons has a son and a daughter and the other son has two sons and a daughter but none of these children have the disease. Mr. H does not have the disease. In taking the family history, Mrs. H tells you that her mother and her only sibling, a brother, have the disease as well, but her niece and nephew do not. She also believes that her mother's mother had the disease but she was not formally diagnosed with it before her death. Some affected members of the family have more severe disease than others but all affected persons show some signs of the disease.

110. The risk for Mrs. H's daughter to have another affected child with this genetic disease is which of the following?

(A) 0%
(B) 25%
(C) 50%
(D) 100%

111. The risk for a son of Mrs. H to have an affected child with this genetic disease is which of the following?

(A) 0%
(B) 25%
(C) 50%
(D) 100%

112. The difference in severity of clinical symptoms among the affected family members in this family may be explained by which of the following?

(A) variable expressivity
(B) allelic heterogeneity
(C) heteroplasmy
(D) incomplete penetrance

Questions 113 and 114 are based on the following family history.

Classical phenylketonuria (PKU) is an important inborn error of metabolism caused by a defect in the enzyme phenylalanine hydroxylase (PAH). PKU causes mental retardation by disrupting myelin formation and protein synthesis in the brain. PKU can be successfully treated by dietary therapy. The gene for PAH maps to chromosome l2q24. Mr. and Mrs. B have two children with PKU, a son and a daughter. Mr. and Mrs. B also have a son and daughter who do not have PKU. No one else in the family has PKU but Mr. and Mrs. B have recently learned that they are third cousins. Mrs. B has a sister who has a daughter and Mr. B has a brother who has a son. They are of Swedish ancestry on both sides of the family.

113. What is the risk that Mr. and Mrs. B could have another affected child with PKU?

(A) 0%
(B) 25%
(C) 50%
(D) 75%
(E) 100%

114. What is the risk that Mrs. B's sister is a carrier of a PKU mutation?

(A) 0%
(B) 25%
(C) 50%
(D) 75%
(E) 100%

Questions 115 and 116 are based on the following family history.

Felicity and Ben have two girls (Angela and Renee) and are expecting a baby boy. Both Felicity and her mother Pam have neurofibromatosis (NF) type I. Angela and Renee have not shown any signs of NF-I to date. Felicity has numerous cafe-au-lait spots (more than 100), neurofibromas, Lisch nodules, axillary and inguinal freckling, and a large plexiform neurofibroma on her back. Pam has 10 small cafe-au-lait spots, 3 neurofibromas, and axillary freckling.

115. What is the risk for Felicity's male child to have NF-I?

(A) 0%
(B) 25%
(C) 33%
(D) 50%
(E) 100%

116. The difference in severity of clinical symptoms between Felicity and her mother can best be explained by which one of the following?

(A) pleiotropy
(B) incomplete penetrance
(C) variable expressivity
(D) anticipation

117. Hemochromatosis is an autosomal recessive disease that is relatively common in the population (1 in 500). The frequency of the mutated gene is thus 0.045. The disease can be treated successfully by periodic removal of blood but failure to recognize it can lead to a number of serious conditions. Population screening for carriers is being considered. What is the expected carrier frequency in the population?

(A) 0.086
(B) 0.100
(C) 0.356
(D) 0.492
(E) 0.706

118. The species of mosquito that carries the malaria parasite *Plasmodium falciparum* is spreading northward, presumably because of global warming, and the incidence of malaria in these regions is on the rise. Which one of the following is the most likely consequence of this range extension?

(A) The frequency of the sickle cell gene (S gene) will decrease in human populations in these areas.
(B) The frequency of people homozygous for the normal allele of the sickle cell gene (A) will increase.

(C) Human A and S gene frequencies will not change.
(D) The frequency of heterozygotes (AS) in the human population will increase.

119. The gene frequency "p" for the autosomal dominant disease, Marfan syndrome, is 0.00005. What is the frequency of "q"?

(A) 0.000707
(B) 0.00025
(C) 0.00025
(D) 0.99995

120. The Philadelphia chromosome is produced by which one of the following?

(A) A balanced reciprocal translocation between chromosomes 9 and 22.
(B) A balanced reciprocal translocation between chromosomes 8 and 21.
(C) A fusion of the p53 and abl proto-oncogenes.
(D) Loss of heterozygosity.

121. In almost every tumor examined, the telomerase gene has been reactivated. This observation may explain which of the following?

(A) how telomeres assist in DNA synthesis
(B) how tumor cells metastasize
(C) how tumor cells become "immortal"
(D) why telomeres are translocated in all cancers

122. Which one of the following chromosomes is never involved in a Robertsonian translocation?

(A) 13
(B) 18
(C) 21
(D) 22

123. At the end of meiosis I, how many chromosomes are present in the daughter cells?

(A) 23
(B) 46
(C) 69
(D) 92

124. A non-disjunction of an X chromosome occurs in a 46,XX conceptus during the first cell division resulting in mosaicism. Providing the embryo goes to term, what cell lines would you expect to be present in the newborn?

(A) 46,XX and 47,XXX
(B) 45,X and 46,XX
(C) 45,X and 47,XXX
(D) 45,X, 46,XX and 47,XXX

125. Bloom syndrome is an autosomal recessive disorder that is associated with a high frequency of spontaneous chromosome breakage due to a defect in a gene involved in DNA replication. There is an increased risk of malignancy associated with this syndrome. The most likely explanation for this observation is which one of the following?

(A) The mutated gene involved is a promoter.
(B) It is more likely that a chromosome rearrangement may occur that will activate a proto-oncogene.
(C) The mutation involved is a frame shift mutation.
(D) There is less cell death (apoptosis) with this syndrome.

Questions 126 and 127 are based on the following family history.

Mr. Z is a 20-year-old man found to have thousands of polyps lining the entire colon. In-vitro synthesized protein assay (IVSP) confirms that the man has a mutation in the APC gene responsible for familial adenomatous polyposis (FAP). His father died of colon cancer at age 40. Mr. Z has a 12-year-old sister, Miss Z.

126. What is the best plan of care for Miss Z?

(A) colectomy
(B) test Miss Z with the IVSP assay to see if she inherited an APC mutation
(C) sigmoidoscopy every 3 months
(D) do nothing besides routine well-child visits

127. Mr. Z's wife is pregnant. What is the chance that his child will inherit the mutation responsible for FAP?

(A) 0%
(B) 25%
(C) 50%
(D) 100%

128. Which one of the following applies to classic phenylketonuria (PKU)?

(A) It is a very rare disease in Caucasians.
(B) It can be treated by limiting phenylalanine intake.
(C) It causes disease only in childhood.
(D) It does not affect the brain.

129. Fluorescent in situ hybridization (FISH) is often used to monitor patients with chronic myelogenous leukemia (CML) for minimal residual disease (MRD) after treatment. The abl oncogene DNA probe is labeled with a red fluorescent molecule and the bcr gene probe is labeled with a green. The bcr/abl fusion gene appears yellow under UV light. Although metaphases are usually available for FISH analysis, interphases may give a better indication of MRD. You "score" a CML patient slide using interphase FISH to detect MRD. Out of the 200 cells you examine, you find that 150 have two green and two red signals and 50 have one green, one red, and one yellow signal. This result indicates which one of the following?

(A) There is no sign of MRD.
(B) The fusion product is not being produced.
(C) At least 25% of the cells examined are malignant.
(D) The sample was not informative.

130. A change in the nucleotide sequence of a gene is which one of the following?

(A) genomic imprinting
(B) transcription
(C) mutation
(D) synthesis

131. A couple, both of whom have a diagnosis of achondroplastic dwarfism, want to know what the risk is of having a child with normal height (they would like another child with dwarfism). Achondroplastic dwarfism is autosomal dominant and the homozygous dominant genotype is lethal very early in development. What is the risk of having a child of normal height?

(A) 0%
(B) 33%
(C) 50%
(D) 100%

Questions 132 and 133 are based on the following pedigree.

This is the family pedigree of Abby who has cystic fibrosis, an autosomal recessive disease.

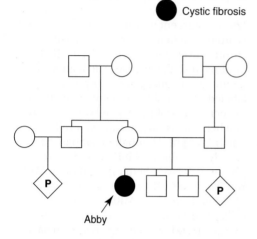

● Cystic fibrosis

Abby

132. What is the risk that any child that Abby may have will be a carrier of a cystic fibrosis mutation?

(A) 0%
(B) 25%
(C) 50%
(D) 100%

133. What is the chance that Abby's brothers are carriers of the cystic fibrosis mutation?

(A) 0%
(B) 50%

(C) 67%
(D) 100%

134. A 25-year-old mother of two is concerned because her 6-year-old son is "slow." She tells you that the father is also slow, as was his father. Neither one finished school, although they both work. The child has no outstanding physical findings. The mother says the child resembles the father. Your assessment of the child is that there is some degree of mental retardation. Assuming that the grandfather, father, and son have the same genetic basis for their mental retardation, the pedigree for this family would indicate which of the following?

(A) autosomal recessive inheritance
(B) autosomal dominant inheritance
(C) X-linked recessive inheritance
(D) multifactorial inheritance

135. In the Pingelapese population of the eastern Caroline Islands in the Pacific, a severe ocular abnormality characterized by total colorblindness, photophobia, and cataract development is found in approximately 10% of the population. Inheritance is autosomal recessive. In 1780, a typhoon struck the island, killing most of the inhabitants. The survivors included only nine males. The general population incidence of this disorder is 1 in 100,000. The high incidence of this disorder in the Pingelapese population is most likely due to which of the following?

(A) assortative mating
(B) selection against heterozygotes
(C) founder effect
(D) eugenics

Answers and Explanations

1. **The answer is (C).** Diabetes and most cancers have a number of environmental and genetic components that are involved in their causation. Marfan syndrome and neurofibromatosis are autosomal dominant disorders, Klinefelter and Turner syndromes are caused by aneuploidy of the sex chromosomes, and cystic fibrosis and sickle cell anemia are autosomal recessive diseases.

2. **The answer is (A).** Human papilloma viruses (HPV) have been implicated as the chief cause of cervical cancer. BCR, BRCA1, and Rb1 are genes involved in the development of various cancers.

3. **The answer is (A).** There is a risk for Robertsonian Down syndrome with this translocation. All the other combinations would be lethal if Robertsonian trisomies resulted, spontaneously aborting very early in the embryonic period.

4. **The answer is (C).** Carriers of balanced chromosome rearrangements, such as reciprocal translocations, are at risk for having conceptions with unbalanced rearrangements that can lead to miscarriage.

5. **The answer is (B).** In Fragile X syndrome, the CGG triplet repeat is sometimes expanded as it is passed on through carrier females and once it reaches 230 copies in an individual, that individual is affected. Males can pass on a "premutation" with fewer copies and there is no expansion when males pass it on. Females who carry expansions over 230 may be affected due to skewed X inactivation.

6. **The answer is (C).** In leukemia, remission is marked by the disappearance of the chromosome abnormality associated with the particular type of leukemia. The derivative chromosome 22 (the Philadelphia chromosome) associated with CML disappears in remission.

7. **The answer is (B).** An affected individual in the family is the ideal person to be tested because the specific mutation being passed on in the family can be identified by testing that individual. Other individuals in the family can then be tested for that specific mutation with accuracy and a considerable reduction in the cost of testing. Lila is the only affected individual in the family still living, so she is the ideal person in the family to be tested.

8. **The answer is (A).** BRCA mutations are inherited in an autosomal dominant manner and confer a 50%–80% lifetime risk for breast cancer depending on the specific mutation involved. Preventative options include monitoring by mammography. Risks of BRCA testing include psychological distress for positive AND negative results (guilt for not having the mutation) and this can change relationships between positive and negative members of a family. Even with a negative result, women still have a 1 in 8 lifetime risk of breast cancer from other causes.

9. In hematological cancers, structural chromosome rearrangements such as translocations activate proto-oncogenes by juxtaposing them.

10. **The answer is (A).** FAP confers a 100% lifetime risk of colon cancer if the polyps that develop into the cancer are not removed.

11. **The answer is (C).** The BRCA1 and 2 mutations convey an approximate lifetime risk of ovarian cancer of 40% and a 6% risk of breast cancer for males in the family.

12. **The answer is (A).** The tandem mass spectroscopy test for PKU is inexpensive, accurate, and reliable. PKU is serious but an effective treatment, diet, is available and there are

numerous treatment centers. The screening tests for the other disorders are expensive, treatment success varies, and resources for diagnosis and treatment are not always readily available.

13. **The answer is (B).** Autism is a component of a number of chromosome abnormalities, including Fragile X syndrome, but is a multifactorial disorder with a low recurrence risk. A link between autism and childhood vaccines or gluten in the diet has not been convincingly established.

14. **The answer is (A).** In hereditary cancer syndromes, the inheritance of a mutated gene is the first "hit" in the sequence of events leading the development of the cancer so the risk of second hits is increased, thus leading to the increased risk for bilateral or multiple tumors.

15. **The answer is (A).** Because a strong genetic component to a trait would have a concordance approaching 1.0 in monozygotic twins and have a large difference between the concordance of monozygotic and dizygotic twins, concordance that is almost equal between monozygotic and dizygotic twins means that the environmental component is stronger than the genetic component.

16. **The answer is (A).** Type 2 diabetes is far more common than type 1 diabetes, it is multifactorial with both genetic and environmental components, and recurrence risk in families is quite high. Concordance in monozygous twins is approximately 90%. There is a strong environmental component that is diet related.

17. **The answer is (C).** The risk of developing cancer in adenomatous polyposis coli is 100% if polyps are not removed. If they are removed, the cancer can effectively be prevented. Prevention is not as successful with BRCA mutations, and is not possible for Huntington disease or cystic fibrosis.

18. **The answer is (A).** All 50 states screen for PKU, galactosemia, and hypothyroidism. Other screening tests offered in the United States vary from state to state.

19. **The answer is (D).** In X-linked recessive disorders, the incidence of the disease is the incidence of the mutated gene "q." The frequency of carrier females is "2pq." Because "q" is so small, "p" is close to 1, so the frequency of carrier females is $2(1)(1/20,000)$ $= 2/20,000 = 1/10,000$.

20. **The answer is (B).** Autism is a component of many chromosome abnormalities so cytogenetic testing should always be done when autism is diagnosed in a patient.

21. **The answer is (D).** In genetic lethals, the disease gene cannot be passed on, so the presence of the gene in the population is entirely due to new mutations that recreate the disease gene.

22. **The answer is (D).** Sickle cell disease is autosomal recessive, Huntington disease is autosomal dominant, Rett syndrome is X-linked dominant, and Duchenne muscular dystrophy and Fragile X syndrome are X-linked recessive.

23. **The answer is (E).** Fragile X syndrome is due to expansion of a triplet repeat, CGG. Sickle cell disease is autosomal recessive, Huntington disease is autosomal dominant, Rett syndrome is X-linked dominant, and Duchenne muscular dystrophy and Fragile X syndrome are X-linked recessive.

24. **The answer is (B).** Rett syndrome is generally lethal in males and is characterized by mental retardation and characteristic hand movements in affected females, among other findings. Mental retardation is associated with Fragile X syndrome, but inheritance is autosomal recessive and mostly males are affected. Mental retardation is not associated with Huntington disease, Duchenne muscular dystrophy, or sickle cell disease.

25. **The answer is (C).** These findings are associated with the degeneration of muscle tissue characteristic of Duchenne muscular dystrophy and are not associated with any of the other conditions listed.

26. **The answer is (E).** Fragile X syndrome is the most common cause of inherited mental retardation. Mental retardation is a finding in individuals with Rett syndrome. None of the other disorders listed are associated with mental retardation.

27. **The answer is (C).** The 22q11.2 deletion has a 50% chance of being passed on because there is a 50% chance of a gamete receiving the maternal or paternal chromosome 22 during meiosis. Because Hillary's father has a heart defect, it is likely that he also has the 22q11.2 microdeletion and passed it on to his daughter.

28. **The answer is (C).** Chromosome abnormalities are often spontaneous in origin. In autosomal dominant and X-linked dominant disease, there would most likely be other affected family members and in mitochondrial disease, all children of an affected mother would be affected.

29. **The answer is (D).** Cri-du-chat (5p-) syndrome is a cytogenetically visible deletion so FISH is not necessary.

30. **The answer is (D).** Although infants with Tay-Sachs disease show no symptoms initially, they begin to display symptoms sometime around six months of age and steadily deteriorate until death at approximately 5 years of age. Managing phenylalanine intake is important for pregnant women with PKU, carriers of sickle cell trait are generally asymptomatic, and there are many cystic fibrosis mutations that are not included in the common screening tests.

31. **The answer is (A).** Empirical data (Table) shows that the recurrence risk for an isolated sacral appendage in North Carolina with one sibling affected is 5%.

32. **The answer is (C).** The risk to each pregnancy in an autosomal dominant disorder is 50%.

33. **The answer is (A).** Because cystic fibrosis is an autosomal recessive disease, I-1 is an obligate carrier.

34. **The answer is (B).** For any pregnancy with an autosomal recessive disorder, there is a 1/4 chance of not being a carrier, a 1/2 chance of being a carrier, and a 1/4 chance of being affected. Because II-2 does not have cystic fibrosis there is a 1/3 chance of the remaining possibilities that she is not a carrier, but a 2/3 chance of the remaining possibilities that she is.

35. **The answer is (D).** For any pregnancy with an autosomal recessive disorder, there is a 1/4 chance of not being a carrier, a 1/2 chance of being a carrier, and a 1/4 chance of being affected.

36. **The answer is (D).** Ultrasonography is usually performed, other than "dating" at around 12 weeks, in the period from 15 to 18 weeks of gestation and is approximately 80% accurate in detecting major structural abnormalities. The "banana sign" and the "lemon sign" are associated with myelomeningocele.

37. **The answer is (A).** CVS and amniocentesis are the definitive tests for the prenatal diagnosis of chromosome abnormalities. CVS has a higher risk for miscarriage than amniocentesis, and amniotic fluid AFP can be assayed to rule out neural tube defects.

38. **The answer is (D).** Because hemochromatosis is an autosomal recessive disease, the mother is an obligate carrier.

39. **The answer is (A).** The risk of a chromosome abnormality for women who will deliver at age 35 is >1% and will increase every year until menopause. Prenatal diagnosis would be indicated for a male fetus if there was a known X-linked disease in the family. Fourth and fifth relatives are so distantly related that any risks associated with genetic disorders are close to the population risk.

40. **The answer is (A).** Because males cannot pass on an X chromosome to their sons, in an X-linked dominant condition all the daughters would be affected because the father only had the one X with the mutation to pass on.

41. **The answer is (C).** In an X-linked recessive condition, males will be affected and carrier females will not. Because sons do not receive an X chromosome from their father, only daughters have a chance of becoming carriers.

42. **The answer is (B).** Autosomal recessive conditions generally appear in a family without a prior family history. The parents are unaffected obligate carriers.

43. **The answer is (A).** In autosomal dominant mutations, there is a 50% chance of passing on the mutation. Autosomal recessive inheritance is not likely since these conditions are rare so the chance that the mother is a carrier of an autosomal recessive mutation is very small. Because a father cannot pass on an X chromosome to his sons, passing on X-linked conditions to the sons is not possible.

44. **The answer is (B).** It is not likely that daughters would be affected with X-linked recessive disorders. Autosomal and X-linked dominant disorders can affect both sexes in any generation. All children of a mother with a mitochondrial condition would be affected.

45. **The answer is (A).** When multiple siblings are affected with an autosomal dominant or X-linked disease and the parents are unaffected with no family history, germline mosaicism is the likely explanation. Mutations of different genes that cause the same phenotype is locus heterogeneity, an affected individual having more severe disease than a sibling is an example of variable expression, and penetrance is an all or none phenomenon where an individual may or may not express a phenotype, even though the disease genotype is present.

46. **The answer is (B).** Amniocentesis is typically done from 15 to 18 weeks of gestation and has a very low risk of complications.

47. **The answer is (C).** Taking a blood sample from the umbilical cord allows management of Rh incompatibility as well as the diagnosis of hematological abnormalities.

48. **The answer is (B).** CVS is typically done earlier than amniocentesis, at 10–12 weeks. However, it carries a larger risk for complications, most notably limb defects.

49. **The answer is (C).** An AFP assay can be performed on amniotic fluid to determine if the AFP is elevated. This could be an indication that a neural tube defect is present. This test cannot be performed with CVS. Cordocentesis is useful for assessing hematological status or infection and the blood obtained can be used for karyotyping. Screening tests are useful for screening at risk populations but the risk is not eliminated if the test is negative.

50. **The answer is (E).** When genes producing the same phenotype are found at different loci, this is called locus heterogeneity. These genes are generally not linked. Synteny refers to genes on the same chromosome, which do not usually cause the same phenotype and are usually not linked. Allelic heterogeneity refers to the situation when different mutations at the same locus cause the same phenotype.

51. **The answer is (B).** These findings are the hallmark of Tay-Sachs disease, which is a degradation pathway disorder. The disease progresses from normal at birth to an early death because of the buildup of a ganglioside in neurons due to the inability of a lysosomal enzyme to degrade it.

52. **The answer is (D).** Metabolic diseases are usually due to the failure of an enzyme in a biochemical pathway and are not associated with growth disorders. All the other choices are clinical findings in metabolic disorders.

53. **The answer is (C).** The findings described with the history are generally seen in metabolic disorders, which are not associated with cytogenetic abnormalities. Performing a chromosome study would not be helpful in this instance.

54. **The answer is (C).** Because the Prader-Willi region of chromosome 15 is imprinted, uniparental disomy for the maternal chromosome 15 would mean that there are two copies

of the maternal region of chromosome 15 but no copies of the paternal region resulting in Prader-Willi syndrome.

55. **The answer is (D).** More severe disease with an earlier age of onset in succeeding generations is called anticipation. Anticipation can also be observed in diseases with triplet repeat expansions such as Fragile X syndrome.

56. **The answer is (D).** All cancers have a genetic component but most cancers are multifactorial diseases. Only a small group of cancers are inherited. Chromosome abnormalities of all kinds, not just translocations, are associated with many hematological disorders and solid tumors, but in many cases, chromosome aberrations are related to progression of the cancer and not necessarily the cause.

57. **The answer is (A).** Retinoblastoma is inherited in an autosomal dominant fashion. The inherited mutated tumor suppressor gene is considered the first "hit" in the progression to a malignant tumor but it takes a mutation in the other allele for the cancer to occur. So, as in the case of autosomal recessive diseases, both mutated alleles must be present before the disease occurs, but in retinoblastoma, the second mutation is acquired, not inherited.

58. **The answer is (B).** Even though a diagnosis of trisomy 21 can often be made based on the phenotype alone, cytogenetic analysis can confirm the diagnosis in those cases as well as the ones where the diagnosis may not be as obvious. It is also important to rule out the possibility of Down syndrome due to a Robertsonian translocation, as this may have reproductive ramifications for the family.

59. **The answer is (C).** Humans have approximately 30,000 genes versus around 19,000 for *C. elegans* but have the most non-coding DNA of any organism examined to date. Humans have 46 chromosomes but one species of tree frog has 48 chromosomes and carp have 100 chromosomes.

60. **The answer is (B).** In the autosomal dominant Huntington disease, both homozygotes and heterozygotes will have the disease and there is no difference in clinical course with either genotype. In small populations founded by a few individuals, some genotypes may be over-represented than would be expected by chance. If individuals with the Huntington gene had more descendants than those with other genotypes, the frequency of the Huntington gene would be expected to increase in the population.

61. **The answer is (D).** In X-linked recessive disorders, the frequency of the gene is the same as the frequency of the disease. Because the disease frequency is 1/20,000, this equals 0.00005 and that is the gene frequency as well.

62. **The answer is (B).** If a balanced reciprocal translocation involves very large segments of the chromosomes, unbalanced outcomes in a conception may be lethal. Sometimes development ends so soon after conception that the couple may not even know that they were pregnant. Perceived infertility would result. There is still a chance of having normal children and the risk for an abnormal child with this couple seems to be small. If they have managed to conceive, none of the unbalanced conceptions have apparently developed much beyond conception.

63. **The answer is (C).** A group of birth defects with a single cause is a syndrome. In an association, two or more traits occur together more often than by chance but do not necessary have the same cause. In a deformation, the normal development of a body part is altered by mechanical forces. A disruption is a defect in a morphological structure due to the failure of a normal developmental process.

64. **The answer is (A).** Allelic heterogeneity can cause the same disease or clinically distinct diseases by different mutations at the same locus.

65. **The answer is (B).** Because the Prader-Willi/Angelman locus is imprinted, a deletion of that region on the paternal chromosome 15 will cause Prader-Willi syndrome and a deletion of that region on the maternal chromosome 15 would cause Angelman syndrome.

66. **The answer is (A).** There is a 50% or 1/2 chance that your patient inherited the mutation, a 50% or 1/2 chance that she will pass it on, and a 50% or 1/2 chance that the child will be a male and thus affected. So, $1/2 \times 1/2 \times 1/2 = 1/8$ chance to have an affected fetus.

67. **The answer is (A).** Low AFP, high HCG, and low uE3 values in a triple screen test indicates an elevated risk for Down syndrome. Amniocentesis with karyotyping is a very low risk, extremely accurate test for detecting Down syndrome.

68. **The answer is (B).** FISH is the definitive test for microdeletions. Klinefelter and Turner syndromes can be detected with conventional chromosome analysis. FISH is not the test of choice for the other disorders listed.

69. **The answer is (A).** A fetus with anencephaly is not viable. Neural tube defects are multi-factorial and not associated with the mother's age at conception. A high maternal serum AFP indicates an increased risk for a fetal open neural tube defect.

70. **The answer is (B).** In meiosis, inverted chromosomes have to form a loop to pair up with the normal homolog. If crossing over occurs in the inversion loop, unbalanced recombinant chromosomes with duplications and deletions can result.

71. **The answer is (C).** Approximately 90% of 45,X conceptions spontaneously abort. Those that survive to birth are females with Turner syndrome. Individuals with Turner syndrome do not have a male phenotype, they are not mentally retarded, and there is no association between Down syndrome and Turner syndrome.

72. **The answer is (A).** When mutations at different loci result in the same phenotype, that constitutes locus heterogeneity. Allelic heterogeneity occurs when mutations at the same locus produce the same disease or clinically distinct diseases.

73. **The answer is (A).** The 9;22 translocation fuses the ABL and BCR proto-oncogenes, creating a chimeric or fusion gene in which both proto-oncogenes are activated.

74. **The answer is (E).** A brother is a first-degree relative. First cousins and great-grandchildren are third-degree relatives. A second cousin would be a fourth-degree relative. Half-siblings are second-degree relatives.

75. **The answer is (B).** The risk that Mrs. X's mother inherited the mutation is 50% or 1/2. The risk that Mrs. X inherited the mutation from her mother is 50% or 1/2. The chance that Mrs. X will pass on the mutation to the fetus is 50% or 1/2. The risk that the male fetus will have the mutation is 50%. So the risk is $1/2 \times 1/2 \times 1/2 = 1/8$ that the fetus will have hemophilia A.

76. **The answer is (B).** Amniocentesis is the test of choice for detecting, through chromosome analysis, Down syndrome. The triple screen is generally done outside the optimal time for CVS and CVS carries a higher risk for complications. Ultrasonography can pick up some signs of Down syndrome but is not as accurate as chromosome analysis. PGD does not apply as the patient is already pregnant.

77. **The answer is (C).** In an autosomal dominant syndrome, a first-degree relative would have a 50% chance of inheriting the mutation, second-degree relatives a 50% chance of the 50% risk the first-degree relative has or 25%, and a third-degree relative would have a 50% chance of the 25% risk the second-degree relative has or 12.5%.

78. **The answer is (C).** In an X-linked recessive disorder, the possible outcomes would be: a 1/4 chance of having a female who is a carrier, a 1/4 chance of having a female who is not a carrier, a 1/4 chance of having a male who is unaffected, and a 1/4 chance of having a male who is affected.

79. **The answer is (A).** The number of individuals who are truly affected is the proportion of people in the population who have the null mutation, 0.003, so $0.003 \times 20,000 = 60$ people in the population have the null mutation. Because the test has a sensitivity of 95%, it will detect 95% of those who are truly positive, or $0.95 \times 60 = 57$ of the people who are

really affected will be detected by the test. The test has a specificity of 99%, so it will identify 99% of those who are negative for the null mutation. The remainder, $0.1 \times 20,000 = 200$ people will be identified as positive for the null mutation. So the number of people the test will indicate are positive are the 57 the test correctly identifies as positive plus the 200 people the test indicates are positive, or $200 + 57 = 257$ people the test would call positive.

80. **The answer is (C).** Since MSAFP is offered in the second trimester, it is generally done after the optimal time for chorionic villus sampling (CVS). An adjustment of seven days in determining gestational age would not appreciably affect the MSAFP results. A woman who delivers at age 35 or older is at an increased risk to have a child with Down syndrome and other chromosome abnormalities and negative MSAFP results would not eliminate that risk. A high MSAFP may indicate that there is an open neural tube defect so ultrasonography should be done to look for defects and amniocentesis should be offered to determine if a chromosome abnormality is present.

81. **The answer is (E).** Because the development of multiple cancers in an individual is a rare event, a proband with multiple cancers indicates that there are probably strong genetic factors, consistent with familial cancers, involved in the development of the multiple cancers.

82. **The answer is (D).** In autosomal dominant diseases, homozygotes are rare so most of those affected are heterozygotes. The frequency of the allele "p" is 0.0001. Because "p" is very small, "q" is close to 1 and the frequency of heterozygotes, those affected, is 2pq or $2(0.0001)(1) = 0.0002$.

83. **The answer is (C).** The carrier of a 14;14 translocation has no risk of having an abnormal child due to the translocation because unbalanced conceptions are lethal and do not develop much beyond conception. The pregnancies abort so early that the patient may not know that she is pregnant. This will be perceived as infertility. She is, however, capable of having a normal child or a carrier like herself, who will also be phenotypically normal.

84. **The answer is (D).** Because no mitochondria can be passed on through the sperm, only through the ovum, a father cannot pass on a mitochondrial mutation to any of his children.

85. **The answer is (B).** The recurrence risk in multifactorial disorders is higher when there is a more severe phenotype because the increased severity means that more of the components that cause the phenotype are present.

86. **The answer is (B).** The risk to have an affected child in autosomal recessive disorders when both parents are carriers is 25%. The chance of having a child who is not affected and not a carrier is also 25%. The chance that a child will be an unaffected carrier of the mutation is 50%.

87. **The answer is (B).** In an inversion, there are two breakpoints. The piece between the two breakpoints flips over into the opposite direction, so the genes on that section are in reverse order.

88. **The answer is (D).** An individual with a Robertsonian translocation who has a normal phenotype will have 45 chromosomes because two of the chromosomes, in this case a 13 and a 21, have fused into one chromosome, thus lowering the total count to 45. An individual with a Robertsonian translocation and 46 chromosomes has an additional copy of one of the chromosomes involved in the translocation. In this case, the individual would have an extra copy of either chromosome 13 or 21. If it was a copy of chromosome 13, the individual would have a trisomy 13 phenotype. If the extra copy was chromosome 21, the phenotype would be that of trisomy 21. A 45, X karyotype is found in Turner syndrome. A 47,XXY individual would have Klinefelter syndrome.

89. **The answer is (D).** In the case of a 21;21 Robertsonian translocation carrier, both copies of that individual's chromosome 21 are translocated to each other to form one chromosome.

At meiosis, the translocation 21;21 chromosome can only go to one daughter cell or the other, so at the end of meiosis there will be, in females, either one ovum with two copies of chromosome 21 in the translocated chromosome or an ovum with no copies of chromosome 21. At fertilization, the sperm will donate one copy of chromosome 21 to the conception and if the ovum contains the translocation 21;21 chromosome, there will be three copies of chromosome 21 in the conceptus. If pregnancy is successfully completed, the baby will have Down syndrome. Fertilization of an ovum with no copies of chromosome 21 will not proceed far since this is a lethal condition.

90. **The answer is (C).** Heart disease and diabetes are multifactorial diseases with both genetic and environmental components. Robertsonian translocations are chromosomal events, cystic fibrosis and sickle cell anemia are autosomal recessive diseases, and Duchenne muscular dystrophy and hemophilia are X-linked recessive disorders.

91. **The answer is (D).** Because that individual has only one X, a mutation on that X would cause the disease in that individual. All the other individuals have two X chromosomes, and because Duchenne muscular dystrophy is X-linked recessive, they would be carriers, but not afflicted with the disease.

92. **The answer is (A).** In multifactorial diseases, there is often sex bias where one sex is more frequently affected than the other. Most multifactorial diseases have a recurrence risk of <10%.

93. **The answer is (C).** Penetrance is the percentage of those who actually have a disorder versus those who have the genotype but are not affected. **Because** Jim must have the Marfan genotype but is not affected, then this constitutes non-penetrance for this disorder. Penetrance is all or none. In variable expressivity, a disorder may have degrees of expression but the disorder phenotype is present to some degree. In allelic and locus heterogeneity, a disease phenotype is also present.

94. **The answer is (D).** The homeotic or HOX gene family plays a major role in early development. Proto-oncogenes are involved in some aspects of later development but do not having any role in spatial patterning. The delta F508 mutation is the most common one in cystic fibrosis.

95. **The answer is (A).** The risk of a chromosome abnormality begins to increase in women who 35 or older. Paternal age does not carry an increased risk until age 50 or older and that risk is mostly for single gene mutations. The karyotype of most miscarriages, about 50%, is normal and this does not increase the risk of a chromosome abnormality in subsequent pregnancies. Cystic fibrosis cannot be diagnosed or carrier status established by chromosome analysis.

96. **The answer is (B).** A gene is transcribed, the messenger RNA is translated, and a protein is synthesized.

97. **The answer is (C).** If the inactivation of the normal X is skewed and a large number are inactivated, there may be enough active copies of the mutated X chromosome to cause some symptoms of disease. X-linked recessive diseases generally do not show variable expressivity and are usually fully penetrant.

98. **The answer is (B).** The possible outcomes in an autosomal dominant disease with both parents affected are: 1/4 will be homozygous, 1/2 will be affected, and 1/4 will be normal.

99. **The answer is (A).** In a pericentric inversion, the two breakpoints of the inverted segment are on either side of the centromere, with one in the short arm and one in the long arm of the chromosome. Because the breakpoints are rarely equidistant from the centromere, inverting the piece containing it will place it in a different position on the chromosome.

100. **The answer is (B).** Because heterozygotes have some degree of protection against malaria, they will more likely to survive to reproduce and pass on the sickle cell trait. Those who do not have sickle cell trait are more likely to contract malaria and less likely

to live to reproduce. Those with sickle cell disease will most likely not live long enough to reproduce. Because the heterozygotes will more successful at reproduction, sickle cell trait will be selected for in the population and its frequency will increase.

101. **The answer is (B).** Individuals with PML may have bleeding secondary to disseminated intravascular coagulation and are at high risk for strokes. A rapid cytogenetic diagnosis can definitively diagnose PML so that appropriate preventative treatment can be administered.

102. **The answer is (A).** Meiosis is suspended in prophase I during prenatal life and does not resume until ovulation. At the end of meiosis in females, there is usually only one mature ovum. Two polar bodies contain the genetic material from meiosis I and II cell divisions and these usually degenerate. Crossing over occurs during meiosis in both sexes.

103. **The answer is (B).** Because both non-disjoined homologues of chromosome 21 went to the polar body, there are no copies in the ovum. When a normal sperm joins with the ovum, it will contribute its one copy of chromosome 21, so there will only be one copy of chromosome 21 in the conceptus.

104. **The answer is (A).** Carriers of balanced reciprocal translocations have an overall risk of 5 to 15% to have a child with an unbalanced derivative chromosome depending on the sex of the carrier and features of the translocation itself. The 11;22 translocation described here is not a Robertsonian translocation because it does not involve acrocentric chromosomes.

105. **The answer is (A).** Adjacent segregation of translocation chromosomes in a carrier parent will give rise to unbalanced derivative chromosomes in the gametes. The sister may be a balanced translocation carrier since it is virtually certain that one of the parents is a carrier.

106. **The answer is (D).** The alleles (genes) present at any locus determine the genotype. Most of the rest of the genome is not transcribed and its function is not very well understood at present.

107. **The answer is (B).** Exons are the functional portion of the gene that codes for proteins. Introns may be transcribed but are cut out of the final messenger RNA. Promoters and enhancers influence gene function but are not part of the functional portion of the gene.

108. **The answer is (C).** The risk for the male fetus to be affected is 50%. If the male fetus has inherited the X-linked dominant mutation, then there will be a miscarriage because the condition is lethal prenatally in males. The patient's healthy daughter has no risk to have a child with incontinentia pigmenti as she is not affected with this autosomal dominant condition.

109. **The answer is (C).** The possible outcomes are as follows: there is a 50% risk that a child will be a carrier and a 50% risk that a child will be affected.

110. **The answer is (D).** Given the family history, it is obvious that this is a mitochondrial disease. It is passed on in a matrilineal fashion, so all of the daughter's children will be affected.

111. **The answer is (A).** Because males do not pass on mitochondria via the sperm, there is no risk of passing on a mitochondrial disease to offspring.

112. **The answer is (C).** The number of mitochondria passed on with mutations will vary and that will influence the severity of the disease in some cases. This condition is called heteroplasmy.

113. **The answer is (B).** PKU is an autosomal recessive disease and the possible outcomes are as follows: 25% will have PKU, 50% will be carriers, and 25% will be normal.

114. **The answer is (C).** The possible outcomes are that 50% of the children will be normal, and 50% will be carriers.

115. The answer is (D). NF-1 is an autosomal dominant disease and the risk of having a child with the disease is 50%.

116. The answer is (C). NF-1 exhibits variable expressivity, where some individuals in a family have more severe disease than others.

117. The answer is (A). The frequency of the mutated gene is "q" or 0.045. Because the gene is relatively common, rounding off would introduce some error. So "p" in this case is $1.000 - 0.045$ or 0.955. The carrier frequency, or 2pq, would be $2(0.955)(0.045)$ which would be equal to 0.086 or 8.6%.

118. The answer is (D). Heterozygotes for sickle cell trait are resistant to contracting malaria. This confers a selective advantage on the sickle cell trait because heterozygotes are more likely to survive to reproduce. Those who are not carriers are at risk for contracting the disease and those with sickle cell disease are not as likely to survive to reproduce as those who are normal or are carriers of sickle cell trait. Because heterozygotes have the best chance to leave surviving offspring, the frequency of heterozygote carriers (AS) would be expected to increase.

119. The answer is (D). In any population, for two alleles, the population is made up of those with "p" alleles and those with "q" alleles, in any combination. Therefore the population can be described as $p + q = 1$, where the population equals "1". So, if "p" is found at a frequency of 0.00005 in the population then "q" is $1 - p$ or $1 - 0.00005 = 0.99995$.

120. The answer is (A). A balanced reciprocal translocation between a portion of the long arms of chromosomes 9 and 22 produces the derivative chromosome 22 that was named the "Philadelphia" chromosome because of the city where it was discovered. The translocation is the leukemogenic event in chronic myeloid leukemia (CML) and juxtaposes the ABL and BCR proto-oncogenes to form a fusion gene.

121. The answer is (C). Most human cells have a limited lifespan due to the fact that with each cell division, the telomeric repeats at the ends of the chromosomes get shorter. The telomere repeat sequences at the ends of the chromosomes are not synthesized with the rest of the DNA and eventually they disappear, causing cell death. In stem cells, telomerase is activated to synthesize new telomere sequences with each cell division, conferring a degree of immortality upon them. This seems to happen in most cancers with the tumor cells becoming "immortal" because they no longer have a finite lifespan.

122. The answer is (B). Only the acrocentric chromosomes 13, 14, 15, 21, and 22 are involved in Robertsonian translocations. In a Robertsonian translocation, the small short arms, which contain heterochromatin and the highly repeated ribosomal gene arrays are lost and the chromosomes fuse at the centromeres. This makes one chromosome out of the fused long arms of the two acrocentric chromosomes involved in the translocation. The acrocentric chromosomes and the other chromosomes in the genome can be involved in reciprocal translocations, in which chromosomes exchange pieces.

123. The answer is (A). Meiosis I is the "reduction division" in which the 2n number of chromosomes, 46 in humans, is reduced by half to the "n" number, which is 23 in humans.

124. The answer is (C). In mitosis, there are two copies of each chromosome on the metaphase plate and as the cell starts to divide, the copies separate with one copy going into each of the two new cells. If the two copies do not disjoin and go to the same cell, the other cell will not get a copy. The other homologue separates normally and one copy joins the two copies in one daughter cell while the other homologue becomes the sole copy of that chromosome in the other daughter cell. In this case, the two copies of the X chromosome going to one daughter cell along with one copy from the homologous X chromosome gives one cell three copies for a 47,XXX constitution while the other daughter cell with no copies of the X chromosome receives the other copy of the homologous X chromosome for a 45,X constitution.

125. The answer is (B). Because there is increased chromosome breakage in Bloom syndrome, there are also increased chromosome structural rearrangements. The increased chromosome rearrangements raise the risk that proto-oncogenes will be activated by being juxtaposed through translocation or that tumor suppressor genes will be mutated through deletions or breakage within the genes.

126. **The answer is (B).** Because of the family history, the best course of action is to find out if Miss Z has the mutation. If she does, then sigmoidoscopy will be useful in determining how many polyps are present, and then colectomy may be required to remove affected sections of the colon. If she does not have the mutation, then she has the general population risk for colon cancer and nothing needs to be done except routine medical checkups.

127. **The answer is (C).** Because FAP is an autosomal dominant disease, each child has a 50% risk of inheriting the mutation.

128. **The answer is (B).** PKU is a relatively common disease in Caucasians with an incidence of 1 in 10,000. The treatment consists of limiting phenylalanine in the diet and has to be done throughout an affected individual's lifetime. Failure to treat PKU results in progressive mental retardation due to the buildup of phenylalanine in the brain.

129. **The answer is (C).** With 50 of 200 or 25% of the cells having a yellow signal, this indicates that a bcr/abl fusion has taken place in those cells and that there is residual disease in this patient.

130. **The answer is (C).** A mutation is anything that changes the nucleotide sequence of a gene. Genomic imprinting, transcription, and synthesis do not change the nucleotide sequence of genes.

131. **The answer is (B).** In autosomal dominant disorders, the risk of having a child who is homozygous for the disorder is 1/4 or 25%, the risk of having an affected heterozygote is 1/2 or 50%, and the chance for a normal, unaffected child is 1/4 or 25%. In the case of achondroplasia, however, there is no risk for having a homozygous child because that condition is lethal, so of the three remaining possible outcomes, the chance that the child will be an affected heterozygote is 2 out of 3, or approximately 67% and the chance of having a child of normal height is 1 of 3 or approximately 33%.

132. **The answer is (D).** Because it is unlikely that a reproductive partner would be the carrier of a cystic fibrosis mutation, he would only be able to contribute a normal allele to a conception. Abby can only contribute a mutated allele since both of her alleles have the mutation. So the only possible combination in any conception would be a normal allele and a mutated allele, thus any child born to Abby and her normal partner would be a carrier.

133. **The answer is (C).** In autosomal recessive disorders, there is a 1/4 or 25% chance that a child will not be a carrier or affected a 1/2 or 50% of being a carrier, and a 1/4 or 25% chance of being affected with the disorder. In this family, the brothers are not affected so they only have a risk to be carriers. Of the remaining possibilities then, there is a 1/3 or 33% chance for each of them to be a homozygote and not a carrier, and a 2/3 or 67% risk for each of them to be a carrier.

134. **The answer is (A).** A disorder appearing in every generation that can be passed on from father to son is most likely autosomal dominant. Males can only pass on the Y chromosome to sons, so X-linked inheritance is not a possibility in this case. Autosomal recessive and multifactorial inheritance are generally not multi-generational and have low recurrence risks, so they would not be expected to show up in every generation.

135. **The answer is (C).** In small populations, rare genes may be present in proportions greater than in the general population. If even one surviving male had the mutation, that would constitute more than 10% of the male "population" at that time. It is likely that this gene would remain at a high frequency as the population grew, depending on the reproductive success of those possessing it.

Figure Credits

CHAPTER 1

Figure 1-1 From Dudek RW. *HY Cell and Molecular Biology*, 2nd ed. Baltimore: Lippincott Williams & Wilkins; 2007:84.

Figure 1-2 (A–F) From Dudek RW. *HY Cell and Molecular Biology*, 2nd ed. Baltimore: Lippincott Williams & Wilkins; 2007:91.

CHAPTER 2

Figure 2-1 (A–D) From Dudek RW. *HY Cell and Molecular Biology*, 2nd ed. Baltimore: Lippincott Williams & Wilkins; 2007:49, 50.

Figure 2-2 (A) From Dudek RW. *HY Cell and Molecular Biology*, 2nd ed. Baltimore: Lippincott Williams & Wilkins; 2007:51. **(B)** From Dudek RW. *HY Cell and Molecular Biology*, 2nd ed. Baltimore: Lippincott Williams & Wilkins; 2007:51. **(C)** From Dudek RW. *HY Cell and Molecular Biology*, 2nd ed. Baltimore: Lippincott Williams & Wilkins; 2007:51.

CHAPTER 3

Figure 3-1 (A–C) Modified from Dudek RW. *HY Cell and Molecular Biology*, 2nd ed. Baltimore: Lippincott Williams & Wilkins; 2007:76.

CHAPTER 4

Figure 4-1 (A–E) Modified from Dudek and Fix. *BRS Embryology*, 3rd ed. Baltimore: Lippinctott Williams & Wilkins; 2005:231.

Figure 4-2 (A) From McMillan JA et al. *Oski's Pediatrics: Principles and Practice*, 3rd ed. Baltimore: Lippincott Williams & Wilkins; 1999:2243. **(B)** Courtesy of Dr. Ron Dudek. **(C)** From Dudek RW and Louis TM. *High Yield Gross Anatomy*, 3rd ed. Baltimore: Lippincott Williams & Wilkins; 2008:76. **(D)** From Eisenberg RL. *Clinical Imaging: An Atlas of Differential Diagnosis*, 4th ed. Baltimore: Lippincott Williams & Wilkins; 2003:193. **(E)** McMillan JA et al. *Oski's Pediatrics: Principles and Practice*, 3rd ed. Baltimore: Lippincott Williams & Wilkins; 1999:1828. **(F)** From Swischuk LE. *Imaging of the Newborn, Infant, and Young Child*, 5th ed. Baltimore: Lippincott Williams & Wilkins; 2004:853. **(G)** Damjanov I. *Histopathology: A Color Atlas and Textbook*. Baltimore: Lippincott Williams & Wilkins; 1996:431. **(H)** From McMillan JA et al. *Oski's Pediatrics: Principles and Practice*, 3rd ed. Baltimore: Lippincott Williams & Wilkins; 1999:2004. **(I)** From McMillan JA et al. *Oski's Pediatrics: Principles and Practice,* 3rd ed. Baltimore: Lippincott Williams & Wilkins; 1999:248. **(J)** From Damjanov I. *Histopathology: A Color Atlas and Textbook*. Baltimore: Lippincott Williams & Wilkins; 1996:453. **(K)** From: Damjanov I. *Histopathology: A Color Atlas and Textbook*. Baltimore: Lippincott Williams & Wilkins; 1996:453. **(L)** From Dudek R and Fix J. *BRS Embryology*, 3rd ed. Baltimore: Lippincott Williams & Wilkins; 2005:188. **(M)** From Brandt WE, Helms CA. *Fundamentals of diagnostic radiology*, 2nd ed. Baltimore: Lippincott Williams & Wilkins; 1999:555.

CHAPTER 5

Figure 5-1 (A–D) From Dudek RW. *HY Genetics*. Baltimore: Lippincott Williams & Wilkins; 2008.

Figure 5-2 (A) From McMillian JA, DeAngelis CD, Feigin RD, Warshaw JB. *Oski's Pediatrics: Principles and Practice*, 3rd ed. Baltimore: Lippincott Williams & Wilkins; 1999:2232; no figure number. **(B)** From Rubin R, Strayer DS. *Rubin's Pathology: Clinicopathologic Foundations of Medicine*, 5th ed. Baltimore: Lippincott Williams & Wilkins; 2008:1225. **(C, D, E)** From Westman JA. *Medical Genetics for the Modern Clinician*, 1st ed. Baltimore: Lippincott

Williams & Wilkins; 2006:68. **(F)** From Rubin R and Strayer DS. *Rubin's Pathology*, 5th ed. Baltimore: Lippincott Williams & Wilkins; 2008:1224. **(G)** From Eisenberg RL. *Clinical Imaging; An atlas of differential diagnosis*, 4th ed. Baltimore: Lippincott Williams & Wilkins; 2003:1121. **(H)** From Swischuk LE. *Imaging of the newborn, infant, and young child*, 5th ed. Baltimore: Lippincott Williams & Wilkins; 2004:1097.

CHAPTER 6

Figure 6-1 **(A)** From Dudek RW. *HY Cell and Molecular Biology*, 2nd ed. Baltimore: Lippincott Williams & Wilkins; 2007:94. **(B)** From Dudek RW. *HY Embryology*, 3rd ed. Baltimore: Lippincott Williams & Wilkins; 2007:168.

Figure 6-2 **(A)** From Kirks DR. *Practical Pediatric Imaging: Diagnostic Radiology of Infants and Children*, 3rd ed. Baltimore: Lippincott Williams & Wilkins; 1998:186. **(B)** From Kirks DR. *Practical Pediatric Imaging: Diagnostic Radiology of Infants and Children*, 3rd ed. Baltimore: Lippincott Williams & Wilkins; 1998:187. **(C)** From Rubin R and Strayer DS. *Rubin's Pathology; Clinicopathologic Foundations of Medicine*, 5th ed. Baltimore: Lippincott Williams & Wilkins; 2008:1167. **(D)** From Rubin R and Strayer DS. *Rubin's Pathology; Clinicopathologic Foundations of Medicine*, 5th ed. Baltimore: Lippincott Williams & Wilkins; 2008:1167. **(E)** From Rubin R and Strayer DS. *Rubin's Pathology; Clinicopathologic Foundations of Medicine*, 5th ed. Baltimore: Lippincott Williams & Wilkins; 2008:1167. **(F)** From Rubin R and Strayer DS. *Rubin's Pathology; Clinicopathologic Foundations of Medicine*, 5th ed. Baltimore: Lippincott Williams & Wilkins; 2008:1167.

CHAPTER 9

Figure 9-1 Modified from Dudek RW. *HY Cell and Molecular Biology*, 2nd ed. Baltimore: Lippincott Williams & Wilkins; 2007:127.

Figure 9-2 **(A–B)** Modified from Dudek RW. *HY Cell and Molecular Biology*, 2nd ed. Baltimore: Lippincott Williams & Wilkins; 2007:80–81.

CHAPTER 10

Figure 10-1 **(A)** From Westman JA. *Medical Genetics for the Modern Clinician*, 1st ed. Baltimore: Lippincott Williams & Wilkins; 2006:20. Courtesy of Dr. G. Wenger, Children's Hospital, Columbus, OH. **(B)** From Westman JA. *Medical Genetics for the Modern Clinician*, 1st ed. Baltimore: Lippincott Williams & Wilkins; 2006:25. Courtesy of Dr. G. Wenger, Children's Hospital, Columbus, OH. **(C)** From Sadler TW. *Langman's Medical Embyology*, 10th ed. Baltimore: Lippincott Williams & Wilkins; 2006:21. Courtesy of Dr. Barbara duPont, Greenwood Genetics Center, Greenwood, SC. **(D)** Westman JA. *Medical Genetics for the Modern Clinician*, 1st ed. Baltimore: Lippincott Williams & Wilkins; 2006:20. Courtesy of Dr. G. Wenger, Children's Hospital, Columbus, OH. **(E)** From Sadler TW. *Langman's Medical Embyology*, 10th ed. Baltimore: Lippincott Williams & Wilkins; 2006:21. Courtesy of Dr. Barbara duPont, Greenwood Genetics Center, Greenwood, SC. **(F)** Westman JA. *Medical Genetics for the Modern Clinician*, 1st ed. Baltimore: Lippincott Williams & Wilkins; 2006:20. Courtesy of Dr. Krzysztof Mrozek, The Ohio State University at Columbus, Columbus, OH. **(G)** From Rubin R and Strayer DS. *Rubin's Pathology; Clinicopathologic Foundations of Medicine*, 5th ed. Baltimore: Lippincott Williams & Wilkins; 2008:186. **(H)** From Rubin R and Strayer DS. *Rubin's Pathology: Clinicopathologic Foundations of Medicine*, 5th ed. Baltimore: Lippincott Williams & Wilkins; 2008:186.

CHAPTER 11

Figure 11-1 **(A–D)** From Dudek RW. *HY Embryology*, 3rd ed. Baltimore: Lippincott Williams & Wilkins; 2007:149.

Figure 11-2 **(A)** Modified from Dudek RW. *HY Embryology*, 3rd ed. Baltimore: Lippincott Williams & Wilkins; 2007:161.

Figure 11-3 **(A–J)** From Dudek RW. *HY Embryology*, 3rd ed. Baltimore: Lippincott Williams & Wilkins; 2007:150.

Figure 11-4 From Dudek RW. *HY Embryology*, 3rd ed. Baltimore: Lippincott Williams & Wilkins; 2007:159.

Figure 11-5 From Dudek RW. *HY Embryology*, 3rd ed. Baltimore: Lippincott Williams & Wilkins; 2007:161.

Figure 11-6 From Dudek RW. *HY Embryology*, 3rd ed. Baltimore: Lippincott Williams & Wilkins; 2007:162

CHAPTER 12

Figure 12-1 (A1) From McMillan JA et al. *Oski's Pediatrics*, 3rd ed. Baltimore: Lippincott Williams & Wilkins; 1999:1855. **(A2)** From Damjanov I. *Histopathology: A Color Atlas and Textbook.* Baltimore: Lippincott Williams & Wilkins; 1996:9. **(A3)** From Damjanov I. *Histopathology: A Color Atlas and Textbook.* Baltimore: Lippincott Williams & Wilkins; 1996:9. **(B)** From Damjanov I. *Histopathology: A Color Atlas and Textbook.* Baltimore: Lippincott Williams & Wilkins; 1996:454. **(C)** From Shischuk LE. *Imaging of the Newborn, Infant, and Young Child*, 5th ed. Baltimore: Lippincott Williams & Wilkins; 2004:1097. **(D1)** Rubin R and Strayer DS et al. *Rubin's Pathology*, 5th ed. Baltimore: Lippincott Williams & Wilkins; 2008:409. **(D2)** Rubin R and Strayer DS et al. *Rubin's Pathology*, 5th ed. Baltimore: Lippincott Williams & Wilkins; 2008:409. **(E)** From Rubin R and Strayer DS et al. *Rubin's Pathology*, 5th ed. Baltimore: Lippincott Williams & Wilkins; 2008:654. **(F)** From Damjanov I. *Histopathology: A Color Atlas and Textbook.* Baltimore: Lippincott Williams & Wilkins; 1996:10. **(G)** From McMillan JA et al. *Oski's Pediatrics*, 3rd ed. Baltimore: Lippincott Williams & Wilkins; 1999:2240. **(H)** Sternberg SS et al. *Diagnostic Surgical Pathology; volume 1*, 3rd ed. Baltimore: Lippincott Williams & Wilkins; 1999:692. **(I)** From Damjanov I. *Histopathology: A Color Atlas and Textbook.* Baltimore: Lippincott Williams & Wilkins; 1996:8.

CHAPTER 13

Figure 13-1 (A) From Dudek RW. *HY Histopathology*, 1st ed. Baltimore: Lippincott Williams & Wilkins; 2008:138. **(B)** From Rubin R and Strayer DS. *Rubin's Pathology*, 5th ed. Baltimore: Lippincott Williams & Wilkins; 2008:871.

CHAPTER 15

Figure 15-1 (B, C) From Dudek RW. *HY Genetics*, 1st ed. Baltimore: Lippincott Williams & Wilkins; 2008.

Figure 15-2 (A) From Stevens A and Lowe J. *Human Histology*, 2nd ed. Baltimore: Mosby; 1997:42. **(B)** From McMillan JA et al, eds. *Oski's Pediatrics*, 3rd ed. Philadelphia: Lippincott Williams & Wilkins; 1999:2151. **(C)** From McMillan JA et al, eds. *Oski's Pediatrics*, 3rd ed. Philadelphia: Lippincott Williams & Wilkins; 1999:2143. **(D)** Dudek R and Fix J. *BRS Embryology*, 3rd ed. Baltimore: Lippincott Williams & Wilkins; 2005:178. Original source: McMillan JA et al, eds. *Oski's Pediatrics*, 3rd ed. Philadelphia: Lippincott Williams & Wilkins; 1999:396. Courtesy of M.M. Cohen, Jr., Halifax, Nova Scotia, Canada. **(E)** Dudek R and Fix J. *BRS Embryology*, 3rd ed. Baltimore: Lippincott Williams & Wilkins; 2005:183. Original source: McMillan JA et al, eds. *Oski's Pediatrics*, 3rd ed. Philadelphia:, Lippincott Williams & Wilkins; 1999:2149. **(F)** McMillan JA et al, eds. *Oski's Pediatrics*, 3rd ed. Philadelphia: Lippincott Williams & Wilkins; 1999:2239. **(G)** Dudek R and Fix J. *BRS Embryology*, 3rd ed. Baltimore: Lippincott Williams & Wilkins; 2005:183. Original source: From McKusick VA. *Heritable Disorders of Connective Tissue*, 4th ed. St. Louis: CV Mosby; 1972:67. **(H)** McMillan JA et al, eds. *Oski's Pediatrics*, 3rd ed. Philadelphia: Lippincott Williams & Wilkins; 1999:2251. **(I)** From Swischuk. *Imaging of the Newborn, infant, and young child*, 5th ed. Baltimore: Lippincott Williams & Wilkins; 2004:448. **(J)** Sadler TW. *Langman's Medical Embryology*, 9th ed. Baltimore: Lippincott Williams & Wilkins; 2004:394. Courtesy of Dr. M. Edgerton, University of Virginia, Charlottesville, VA. **(K)** McMillan JA et al. *Oski's Pediatrics Principles and Practice*, 3rd ed. Baltimore: Lippincott Williams & Wilkins; 1999:394.

CHAPTER 16

Figure 16-1 (A) From Weinberg RA. *The Biology of Cancer,* 1st ed. Garland Science; 2007:260. **(B)** From Weinberg RA. *The Biology of Cancer,* 1st ed. Garland Science; 2007:260. **(C)** From Weinberg RA. *The Biology of Cancer,* 1st ed. Garland Science; 2007:397.

Figure 16-2 Modified from Dudek RW. *HY Cell and Molecular Biology,* 2nd ed. Baltimore: Lippincott Williams & Wilkins; 2007:126.

Figure 16-3 Modified from Dudek RW. *HY Cell and Molecular Biology,* 2nd ed. Baltimore: Lippincott Williams & Wilkins; 2007:118.

Figure 16-4 Modified from Dudek RW. *HY Cell and Molecular Biology,* 2nd ed. Baltimore: Lippincott Williams & Wilkins; 2007:121.

Figure 16-5 Modified from Dudek RW. *HY Cell and Molecular Biology,* 2nd ed. Baltimore: Lippincott Williams & Wilkins; 2007:122.

Figure 16-6 (A) Rubin R, Strayer DS et al. *Rubin's Pathology,* 5th ed. Baltimore: Lippincott Williams & Wilkins; 2008:1266. **(B)** Rubin R, Strayer DS et al. *Rubin's Pathology,* 5th ed. Baltimore: Lippincott Williams & Wilkins; 2008:1266. **(C)** Spitz JL. *Genodermatoses: A Full Color Clinical Guide to Genetic Skin Disorders.* Baltimore: Lippincott Williams & Wilkins; 1996:76. Courtesy of Lawrence Gordon, MD, New York. **(D)** From Dudek RW and Louis TM. *High Yield Gross Anatomy,* 3rd ed. Baltimore: Lippincott Williams & Wilkins; 2008:52. **(E)** From Dudek RW. *High Yield Histopathology,* 1st ed. Baltimore: Lippincott Williams & Wilkins; 2008:179. Courtesy of Dr. R.W. Dudek. **(F)** Rubin R, Strayer DS et al. *Rubin's Pathology,* 5th ed. Baltimore; Lippincott Williams & Wilkins; 2008:607.

CHAPTER 18

Figure 18-1 From Dudek RW. *HY Genetics,* 1st ed. Baltimore: Lippincott Williams & Wilkins; 2008.

Index